Sustainable Wildlife Management

Edited by Farzana Khan Perveen

Published in London, United Kingdom

Sustainable Wildlife Management
http://dx.doi.org/10.5772/intechopen.100943
Edited by Farzana Khan Perveen

Contributors
Obote Shakacite, Phillimon Ngandwe, Vincent Nyirenda, Donald Chungu, Gyanranjan Mahalik, Diptiman Sahoo, Gyanesh Dash, K.T.K.G. Ranjan Mohanty, Srinivas Acharya, Ehsan Amiri Ardakani, Monali Priyadarsini Mishra, Abubakar Ali Shidiki, Tessa Medong Rosalie, Donfack Azubjio Ultrich, Blessing M. Munyaradzi Mugaviri, Gilbert H. Moyo, Ezra Pedzisai, Cuthbert Maravanyika, Gatriay Tut Deng, Birtukan Tsegaye Demisse, Gatluak Gatkoth Rolkier, Mulugeta Ruot Kuon, Farzana Khan Perveen, Anzela Khan

First published in London, United Kingdom, 2023 by IntechOpen
IntechOpen is the global imprint of INTECHOPEN LIMITED, registered in England and Wales, registration number: 11086078, 5 Princes Gate Court, London, SW7 2QJ, United Kingdom
Printed in Croatia

British Library Cataloguing-in-Publication Data
A catalogue record for this book is available from the British Library

Additional hard and PDF copies can be obtained from orders@intechopen.com

Sustainable Wildlife Management
Edited by Farzana Khan Perveen
p. cm.

This title is part of the Environmental Sciences Book Series, Volume 4
Topic: Environmental Resilience and Management
Series Editor: J. Kevin Summers
Topic Editor: Jose Navarro-Pedreño

Print ISBN 978-1-80356-629-0
Online ISBN 978-1-80356-630-6
eBook (PDF) ISBN 978-1-80356-631-3
ISSN 2754-6713

IntechOpen Book Series

Environmental Sciences

Volume 4

Aims and Scope of the Series

Scientists have long researched to understand the environment and man's place in it. The search for this knowledge grows in importance as rapid increases in population and economic development intensify humans' stresses on ecosystems. Fortunately, rapid increases in multiple scientific areas are advancing our understanding of environmental sciences. Breakthroughs in computing, molecular biology, ecology, and sustainability science are enhancing our ability to utilize environmental sciences to address real-world problems.

The four topics of this book series - Pollution; Environmental Resilience and Management; Ecosystems and Biodiversity; and Water Science - will address important areas of advancement in the environmental sciences. They will represent an excellent initial grouping of published works on these critical topics.

Meet the Series Editor

J. Kevin Summers is a Senior Research Ecologist at the Environmental Protection Agency's (EPA) Gulf Ecosystem Measurement and Modeling Division. He is currently working with colleagues in the Sustainable and Healthy Communities Program to develop an index of community resilience to natural hazards, an index of human well-being that can be linked to changes in the ecosystem, social and economic services, and a community sustainability tool for communities with populations under 40,000. He leads research efforts for indicator and indices development. Dr. Summers is a systems ecologist and began his career at the EPA in 1989 and has worked in various programs and capacities. This includes leading the National Coastal Assessment in collaboration with the Office of Water which culminated in the award-winning National Coastal Condition Report series (four volumes between 2001 and 2012), and which integrates water quality, sediment quality, habitat, and biological data to assess the ecosystem condition of the United States estuaries. He was acting National Program Director for Ecology for the EPA between 2004 and 2006. He has authored approximately 150 peer-reviewed journal articles, book chapters, and reports and has received many awards for technical accomplishments from the EPA and from outside of the agency. Dr. Summers holds a BA in Zoology and Psychology, an MA in Ecology, and Ph.D. in Systems Ecology/Biology.

Meet the Volume Editor

Dr. Farzana Khan Perveen obtained her MSc in zoology from the University of Karachi, Pakistan, MAS in agriculture from Nagoya University, Japan, and Ph.D. in toxicology from the University of Karachi. She was the Founder Chair of the Department of Zoology and Ex-Controller of Examinations at Shaheed Benazir Bhutto University, and Founder Chair of the Departments of Zoology at Hazara University and Kohat University of Science and Technology. Dr. Khan Perveen has supervised many Ph.D., BS, MSc, and MPhil students in their research. She is the author of 40 books, 9 chapters,150 high-impact research papers, 135 abstracts, and has edited 9 books. She has organized and participated in numerous international and national conferences and received multiple awards and fellowships. She is on the editorial boards of several journals. Dr. Khan Perveen is a member of the World Commission on Protected Areas, the International Union for Conservation of Nature, and a number of research societies. Her fields of interest are agronomy, entomology, and zoology.

Contents

Preface

This book is a concise and comprehensive compendium of knowledge about sustainable wildlife management. Wildlife species are declining and facing many problems worldwide. The major threats to wildlife include poaching, habitat destruction, deforestation, overexploitation, pollution, agriculture and etc. Continuous conservation efforts to tackle these threats are being undertaken by both national governments and international bodies. Among the non-governmental organizations (NGOs) working on these issues are the World Wildlife Fund, Wildlife Protection Society, Wildlife Save Our Souls and Save Our Ship (SOS).

In 1964, the International Union for Conservation of Nature (IUCN) established the Red List, an inventory showing the conservation status of wildlife species as a barometer of life (**Table 1**). Several wildlife species are illustrated in **Figure 1**, with their conservation status according to this categorization.

Chapter 1, "Introductory Chapter: Sustainable Wildlife Management" by Farzana Khan Perveen and Anzela Khan, is an introduction to wildlife management. In many cases, the native fauna of a region is, consciously or unconsciously, limited to vertebrate species, and it sometimes excludes fish. Fish, however, at least fresh water fish, are considered wildlife worldwide, as they are part of the same ecosystems and their management is analogous. Likewise, butterflies and other invertebrates are usually included in wildlife inventories. Wildlife is also used as a term for undomesticated

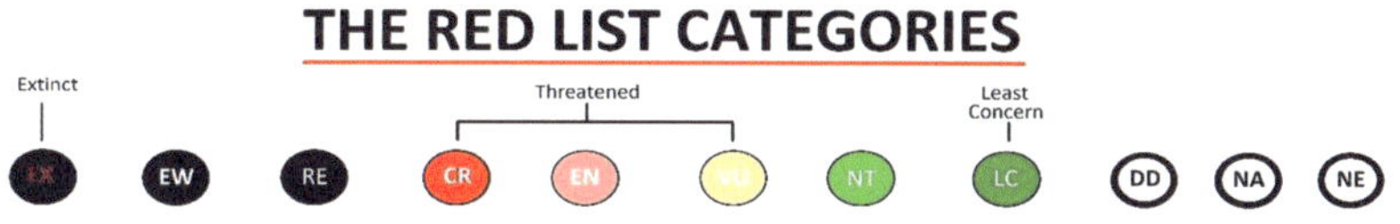

Extinct: EX; Extinct in the Wild: EW; Regionally Extinct : RE; Critically Endangered: CR; Endangered: EN; Vulnerable: VU; Near Threatened: NT; Least Concern: LC; Data Deficient: DD; Not applicable: NA; Not evaluated: NE.

Table 1.
Red List Species categories of the International Union for Conservation of Nature (IUCN).

Figure 1.
a): The hippopotamus, Hippopotamus Amphibius *L., 1758: vulnerable (VU); b): the black rhino,* Diceros bicornis *L., 1758: critically endangered (CE); c): the vaquitas,* Phocoena sinus *Norris and Mc Farland, 1958: critically endangered (CE); d): the kiwis,* Apteryx onii *Gould 1847; e): the megabombus,* Bombus reinigiellus *Rasmont, 1983: uncertain (UC) (IUCN).*

animals and plants that grow independently of people, usually in natural conditions. Traditionally, wildlife includes all game species in the universe, including wild animals hunted for food or sport.

Chapter 2, “Freshwater Fish Migration: Fisheries Management Strategy Insight” by Deng and Demisse, discusses migration, including spawning, feeding, and refuge-seeking, as a deliberate survival strategy by animals. Fish migration activities are influenced by abiotic factors including variations in water temperature, water level, and light availability. Different sensory mechanisms for accurate orientation have been suggested, including the use of sun position, polarized light patterns, and the Earth’s geomagnetic field. Fish morphology plays a significant role in assisting freshwater fish migration. Long-distance migrants have streamlined body structure and longer caudal regions, while short-distance migrants are fusiform, making it hard to move long distances against currents. Since fish migration may involve two different aquatic environments, all migrant fish that cross the interface between freshwater and saline water habitats undergo physiological changes. Protective enclaves are used in wildlife protection and management structures around the world to contain pressure on biodiversity. However, little attention has been paid to species living in unprotected areas, often known as free areas.

Chapter 3, “Status of Hippopotamus, *Hippopotamus Amphibius* L., in the River Sanaga of the Centre Region of Cameroon” by Shidiki et al., discusses the sustainable management of the hippopotamus population in the river Sanaga, facing the threat of extinction due to poaching, fishing, sand mining and crop farming along the river bank. To reduce these threats, sensitization and awareness campaigns are needed.

Chapter 4, “Spatio-Temporal Distribution of the Black Rhino (*Diceros bicornis* L.) in the Midlands Black Rhino Conservancy, Zimbabwe” by Mugaviri et al., describes the use of geographic information system and remote sensing technologies in detecting, mapping and monitoring change in land use/land cover. The results of their study of the spatio-temporal impacts of land use/land cover changes on *D. bicornis* distribution in Midlands Black Rhino Conservancy indicate that bare land increased by over 160%, while woodland decreased by about 46% over the same period.

Chapter 5, “Floristic and Ethnobotanical Study of Indigenous Plants of Ranapur Reserve Forest, Odisha, India” by Sahoo et al., explore selected indigenous plants of the Ranapur reserve forest of Odisha State, India, and their medicinal properties. Their ethnobotanical and floristic survey highlights the identification and use of native flora, and the preservation of natural and cultivated plant species in the hills of Ranapur, Odisha. A field study of diversity around the Ranapur forest identified and documented 143 plant species belonging to 53 families.

In Chapter 6, “Characteristics of the Tropical Hardwood – Ttree Species for Renewable Energy Production in Zambia” by Shakacite et al., the diversity, abundance, and distribution of the 25 most suitable indigenous tropical hardwood tree species for value-added renewable energy production are discussed. The use of these species by local communities remains poorly understood. The study notes the need to manage hardwood supply sustainably, promote lesser-known hardwood tree species, and diversify their use in the timber industry.

The challenges described in Chapter 7, “Management Challenges of Gambella National Park” by Rolkier and Ruot, include both previous settlement, and current settlement and agricultural investment. Institutional changes resulted in a significant staff turnover in Gambella National Park.

This book will be useful for teachers, professors, researchers, scientists, students, naturalists, conservationists, agriculturalists, hunters, biologists, and others that face the challenges associated with wildlife.

The editor is grateful to all the authors for their contributions to this book. Special thanks go to Martina Scerbe, Author Service Manager, and all the staff of IntechOpen publisher. The experiments described in the book comply with the current laws of the country in which they were performed.

Farzana Khan Perveen
The President,
Classes and Event in Science (C.E.S),
Reims, France

Chapter 1

Introductory Chapter: Sustainable Wildlife Management

Farzana Khan Perveen and Anzela Khan

1. Introduction

1.1. Wildlife

The wildlife means the native fauna and sometimes flora of a region, according to the Oxford Advanced Learner's Dictionary. In various circumstances, however, this natural wildlife is, deliberately or automatically, restricted to vertebrate species, and it occasionally eliminates fish, as indirectly obscure in terms of fish and wild flora and fauna civilizations and facilities in the world. Contrariwise, fish, at least fresh-water fish, is considered wildlife worldwide, as they are part of identical bionetworks and their organization is comparable. Similarly, insects and other invertebrates are typically comprised in wildlife records in a little case in many countries. On the other hand, wildlife is also implied as a word for undomesticated animals alive in the barren, according to the Dictionary of American Heritage and for fauna and flora that raise autonomously of individuals, generally in ordinary environments, according to Thesaurus and Dictionary Advance Cambridge Learner. Traditionally, wildlife includes all game species in the universe, as hunting represents, in the social discourse, a way to approach wilderness. Indeed, according to the Webster's Dictionary, wildlife means wild animals, especially those hunted for food or sport [1].

2. Wildlife classification

The pyramid in **Figure 1** shows the taxonomic position of the wild animals for their description. Out of more than 120,000 species monitored by the International Union for Conservation of Nature (IUCN) Red List of Threatened Species, up to 17,000 have a data-deficient (DD) status [1].

3. Wildlife importance

Wildlife is important for human beings for many aspects. Two such aspects are as follows.

3.1 Religion

Many wildlife species and their secretions and creations may be utilized as holy substances in spiritual sacraments. For instance, hawks, eagles, falcons and their

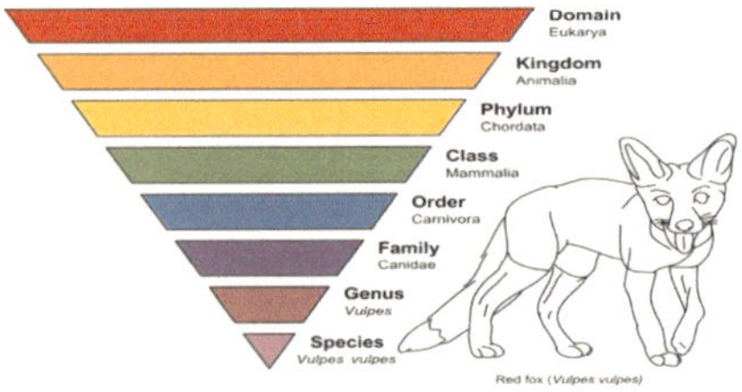

Figure 1.
General criteria for giving the taxonomic position to the wild animal [1].

plumages and fluffs have excessive social and mystical importance to religious objects for the people of the USA. In Hindu religion, the cow is observed as a sanctified animal. Muslims celebrate Eid al-Adha by scarifying animals to memorize Abraham's spiritual scarifications with God's love. Many tetrapod's animals, cows, bulls, goats, sheep and camels may be presented as sacrifice animals during the Eid's three days [2]. The holy book of Christians, that is, the Bible, has names of various animals as symbols of the Lamb as a well-known name of Jesus. In the Novel Testimony, the Gospels, John, Luke and Mark have instinctive cryptograms. The lion is a symbol for Mark, the bull for Luke and the eagle for John [3].

3.2 Tourism

By the wildlife tourism, animals can be vied in their native or similar environments, from vehicles or on foot. Elephants in Hwange National Park, Zimbabwe, were quite undisturbed by people and vehicles (**Figure 2**) [4]. The tourism industry of wildlife is a component of several countries' commerce of travel concentrated everywhere reflection and dealing with resident fauna and flora existing in their usual territories. It comprises human, animal, environment sociable leisure industry and expedition hunting, and is comparable to great event attributions, which come under the canopy of touristic wildlife industry. It, in modest meaning, is networking with non-domesticated creatures in the original habitat, either dynamically (e.g., shooting or gathering) and indirectly (e.g., viewing or cinematography). Wildlife tourism is a significant fragment

Figure 2.
Elephant safari after the one-horned rhinoceros, Rhinoceros unicornis Linnaeus, 1758.

of travel businesses in many nations, for example, African countries, South-American states, Bangladesh, Australia, Canada, India, Indonesia, Malaysia, Maldives, Pakistan and Sri Lanka. This industry has a melodramatic and speedy development in the current global era, and several fundamentals are thoroughly associated with ecotourism and sustainable tourism. According to the United Nations World Tourism Organization (UNWTO), an annual growth of around 3–7% of world tourism industry is related to wildlife tourism. They similarly estimate that the progress is greatly advanced at UNESCO World Heritage Sites (WHS). Around 22 million people worldwide are straightly or circuitously employed in this industry and share additional $120 billion to the total GDP. As a multimillion-dollar international industry, wildlife tourism is frequently categorized by the contribution of modified tour packages and expeditions to permit nearby contact to animals living in natural environment [5] in Royal Chitwan National Park, Manali, India [4].

4. Wildlife biodiversity

Wildlife biodiversity is domineering for several motives; they can be commonly separated into two classes: 1) essential for bioenvironments and 2) importance intended for human beings. Wildlife biodiversity allows ecosystems to grow and become more beautiful. From millions of centuries, countless dissimilar kind of animals and plants have been created to live in identical territories. Over time, they equilibrate independently to each other and grasp the bio-ecosystem assembly. When a type of animal or plant is detached, natural biodiversity is compact, and bio-ecosystem could drop its steadiness, triggering it to break down. One of the great examples is the sea otters, Enhydra lutris (Linnaeus, 1758) in kelp (large brown seed: Laminariales; Agaraceae) forests along the California coast. *Enhydra lutris* feed on sea urchins, *Lytechinus anamesus* Clark, 1912 and *L. anamesus* feed on kelp. If E. lutris are removed, L. anamesus multiply, eating large portions of the kelp forest, destroying the habitat and eventually leading to the death of other animals that live there. The whole ecosystem is destroyed. Wildlife biodiversity is declining at an unprecedented rate. This loss comprises not only genetic, but also ecological and behavioral diversity and is currently not understood clearly. Usually, there are three levels of biodiversity: genetic, species and ecosystem diversity. Genetic diversity is all the different genes contained in all individual animals, plants, fungi and microorganisms [6]. **Table 1** shows the percentage of some threatened with extinction species of wildlife worldwide according to the Red List of IUCN [7].

Figures									
Classes of wildlife	**Mammals**	**Birds**	**Sharks and rays**	**Reptiles**	**Amphibians**	**crustaceans**	**Reef corals**	**Cycads**	**Conifers**
Threatened with extinction	27%	13%	37%	21%	41%	28%	33%	69%	34%

**More than 41,000 wildlife species are threatened with extinction, that is, still 28% of all assessed species.*

Table 1.
Percentage of threatened with extinction species worldwide according to the Red List of IUCN [7].

5. Threats to wildlife biodiversity

Wildlife has faced various pressures in recent decades and has, therefore, attracted attention with regard to research and conservation worldwide. The impacts of human activities or man-made destructions, such as habitat degradation and loss, illegal wildlife trade, introduction of invasive alien species, spread of invasive species and diseases, overexploitation, pollution, changes in disturbance regime, the human impact on the earth's weather-temperature alteration and synergistic features, are the main accountable aspects for the cumulative hazard of extermination for countless wild animal species. Additionally, climatic or weather procedures together with storms, inundations, scarcities, natural or artificial burning of forests and global warming and other several factors are responsible for the extinction of several local-to-wild species. Consequently, suitable maintenance agendas and actions would be functional instantly to avoid more deterioration in the figures or decrease of the topographical circulation of wild fauna and flora. Advances in technology give conservationists, scientists and the general public the advantage to better understand the animals, their habitats and the threats they can face [8]. Wildlife biodiversity has been continually under threat since the dawn of man. The expansion, removal, modification and use of the land are our goals. The changes often damage natural ecosystems and reduce wildlife biodiversity [9].

6. International Union for Conservation of Nature (IUCN)

The International Union for Conservation of Nature (IUCN) had established the Red List since 1964. Then, it has been used as a critical indicator of the health of the world's biodiversity. The Red List can be used as more than just a compendium of species, their status, as a barometer of life an effective tool to inform and inspire actions that may lead to the preservation of biodiversity, which is critical, if want all species to survive. The Red List Index (RLI) shows trends in overall extinction risk for wildlife species and is used by governments to track their progress towards targets for reducing biodiversity loss. According to the Red-List of the IUCN, in 2022, there are 41,415 faunal and floral species; in addition, 16,306 of them are vulnerable fauna and flora endangered with extinction. This number was 16,118 in the previous year. That comprises equally vanishing plants and scarce animals. IUCN's objectives are to influence, encourage and assist societies throughout the world to conserve the integrity and diversity of nature and to ensure that any use of natural resources is equitable and ecologically sustainable. The Red List has various categories of species: extinct, extinct in the wild, critically endangered, endangered, vulnerable, on risk, data deficient and not evaluated (**Table 2**). The IUCN prepared the Red Data Book to keep a record of all endangered animal, plant and fungal species. Through this book, the IUCN is trying to create awareness about the endangered species [7].

6.1 The red list 2022

The Red List prepared by the IUCN every year is widely identified as the most comprehensive source of information on the risk of extinction of animal, plant and fungal species around the world. The world most critically endangered (CR) wildlife species in 2022 has been given in **Table 3** according to the IUCN Red List and the World Wildlife Fund (WWF) of the United Kingdom [10].

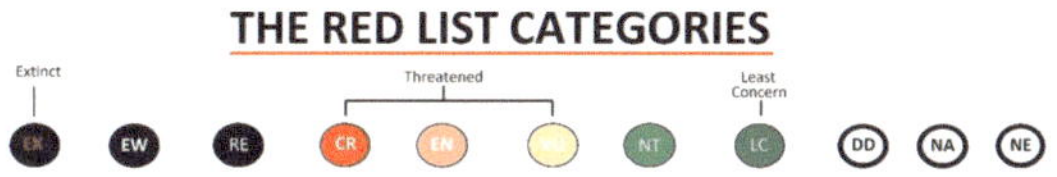

Extinct (EX): no reasonable doubt that the last individual has died. **Extinct in the wild (EW):** known only to survive in captivity, cultivation or well outside its natural range. **Regionally extinct (RE):** disappeared in regional level. **Critically endangered (CR):** facing extremely high risk of extinction in the wild.	**Endangered (EN):** facing a very high risk of extinction in the wild. **Vulnerable (VU):** facing a high risk of extinction in the wild. **Near threatened (NT):** close to qualifying, or likely to qualify for a threatened category in the near future.	**Least concern (LC):** population is stable enough that it is unlikely to face extinction in the near future. **Data deficient (DD):** not enough information on abundance or distribution to estimate its risk of extinction. **Not applicable:** NA **Not evaluated:** NE

Table 2.
The categories of species of the Red List of the International Union for Conservation of Nature (IUCN) [7].

S. No.	Common name	Scientific name with authority	Population	Habitat	Figure
1):	2. AMUR LEOPARD Amur leopard	*Panthera pardus orientalis* (Schlegel, 1857)	around 100	East of Russia and north-eastern China	
2):	Javan rhinoceros	*Rhinoceros sondaicus* Desmarest, 1822	around 60	Tropical forests	
3):	Vaquitas	*Phocoena sinus* Norris and McFarland, 1958	around 10 individuals remain in the wild	only in the northern Gulf of California (Marine)	
4):	Giant manta ray	Manta birostris Bancroft, 1829	around 100 to 1500 individuals	US east coast tropical, subtropical and temperate bodies of water	
5):	Kiwis	*Apteryx onii* Gould 1847	around five individuals living	Okarito forest, New Zealand	
6):	California condor	*Gymnogyps californianus* (Shaw, 1797)	around 44 mature individuals in the wild	Condors to Zion and Grand Canyon National Parks	

S. No.	Common name	Scientific name with authority	Population	Habitat	Figure
7):	Hawksbill turtles	*Eretmochelys imbricata* Linnaeus, 1766	around 20,000–23,000	Nearshore tropical and subtropical of Atlantic, India and Pacific Oceans	
8):	Itatiaia highland frog	*Holoaden bradei* Lutz, 1958	uncertain	Itatiaia Mountains, Brazil	
9):	Megabombus	*Bombus reinigiellus* Rasmont, 1983	uncertain	Spanish Sierra Nevada	
10):	Staghorn coral	*Acropora cervicornis* (Lamarck, 1816)	uncertain	Florida Keys, Bahamas and Caribbean islands Gulf of Mexico	
11):	Forest coconut	*Voanioala gerardii* Dransf., 1989	around 15 mature trees remaining	Madagascar	
12):	Cactus	*Uebelmannia buiningii* Donald, 1921	uncertain	natural habitat, dry savanna, Brazil	

Table 3.
The world most critically endangered (CR) wildlife species in 2022 according to the IUCN Red List and the World Wildlife Fund (WWF) [10].

7. Protection of wildlife biodiversity

There are many ways to protect wildlife biodiversity. Even though the biodiversity of many habitats has become threatened, there are many things to help reduce this danger. These are some of the steps that have been taken to conserve biodiversity worldwide [11].

7.1 Government legislation

Governmental authorities can take actions to regulate the illegal activities that are taken place by smugglers and thieves in the territories inside the countries. Regulation that guards natural habitats via criminalization expansion, reaping of natural resources or additional humanoid mistreatment consumes an enormous power on upholding usual diversity of fauna and flora. Moreover, rules defending explicit biodiversity as in the US Act of Endangered Species (ESA) supports defend biota that have previously been stuck. Caring environments earlier is the best policy to save wildlife. Government can apply the best laws and regulations for wildlife biodiversity conservation. It is efficaciously and strictly executed by related authorities of governments of all the countries [11].

7.2 Nature preserves

Natural preserves are a system of regulation of the government and are frequently recognized as national parks. The regions and organisms are protected by them. As the fauna and flora have been grown by related authorities of governments in the particular places, where general public can be allowed for entrance for amazements with fee. Here, they have the great opportunities to touch or see them very closely. These are outstanding activities for health. The government guards the natural habitat, and the general public can observe the bionetwork. The object mark is that, over time, this assists the society to do supplementary admiration for the natural world and raise compression on administration and management to keep additional areas as natural reserved. Presently, 12% of global territories are reserved and preserved as national parks [12].

7.3 Reduction of invasive species

If a species is not found in an area, it introduces purposely or accidently, it is called as an invasive species. It causes sometimes economic or sometimes environmental destruction. It may be harmful for animals, plants and humans. To minimize it, it is essential to thoroughly check planes, ships and cargo before they are offloaded in a new country. Moreover, it is necessary to consult ecologists, zoologists or botanists of the region for related knowledge before introducing any new species of plants or animals to a particular area [12].

7.4 Restoration of habitat

If an area is destroyed by different human activities, it should be tried to rearrange its natural or original conditions. It means transporting once again the fauna and flora of that area, which were originally occurring here. It has been shown to be a brilliant practise of recurring the plants and animals of a particular area. An example of that is the restoration of wolves, *Canis lupus* Linnaeus, 1758 into National Park Yellowstone. After *C. lupus* come back to the area, they use more elk and coyote as food; it means that the prey species of the coyote are augmented, and river bank areas (riparian) are crushed by elk expand. Those renewal projects can be accepted by NGOs, local organizations and governments [12].

7.5 Seed banks and captive breeding

Great varieties of plant species seeds are stored in the seed's banks. If some species of plants are extinct in the countryside, the seed's banks are furnished these seeds to recover the plant species. When plants are grown by saved seeds, it means that extinct plant species are restored back into its habitation. This is a very serious issue.

Therefore, for many years, seed's banks have been gathering the seeds of more and more plant species. However, there are many seed's banks that have great storage of seeds. Some seed's banks store more than 2 billion seeds. As far as animals are concerned, they are imprisoned during upbringing; in the form of safekeeping, the animals are mostly kept in zoos and farms. This topic seems to be very serious and rather debateable, as it needs to keep animals in capitative form that are frequently nearby extermination. On the other hand, positively, it furnishes the chance to augment the colonies of the animals; hence, they can be re-established into the barren [12].

7.6 Research

To understand the behavior of species, interacting with them in their environment is critical for caring them. As individuals have to supplementary comprehend their communication. To preserve biodiversity, and to aid sheltered the creatures, it is necessary to search novel and supplementary conventional procedures to continue the biodiversity. One case is the usage of biota access strip in developed zones. Research shows that different animals melodramatically increase their populations. It reduces the number of animals that originated into straight joining with people and offers a great range for wandering creatures to transfer extensive reserves [12].

7.7 Reduce climate change

The climate change has disastrous consequences for all living things on earth. Huge amounts of fossil fuels are being used, which directly causes climate change. There is a need to move away from fossil fuels towards alternative energy sources and natural or sustainable products. Reducing the effects of climate change requires a worldwide effort [13].

7.8 Procurement of sustainable products

Today, many products are labeled with ecolabels that state as environmentally sociable. Some of the greatest protruding ecolabels are Rainforest Alliance Certified, Energy Star and USDA Organic. The utilization of ordinary reserves is one of the main motives for biodiversity loss; therefore, it is the duty of the community to consume products that are produced by the maximum justifiable and conceivable means. Moreover, when these goods are consumed, they increase demand for environmentally conscious products pushing producers to produce them more [13].

7.9 Sustainable living

It's enough that it can be chosen to do it on a daily basis. Whether it's driving a motorcycle to work or buying eco-labelled products. Its assistances to decrease the amount of possessions have been used. That is possibly the greatest significant method of protecting biodiversity so that everybody can organize the subject frequently through once minor routine fluctuations. If the whole world selects to live sustainably, then a variety of habitats will improve their biodiversity [13].

7.10 Education

There are many green environmental topics, education is one of the secrets to victory. To provide education to public, the essentiality of biodiversity conservation increases civic consciousness of the problems. As community responsiveness increases, the society develops additional elaborateness and ultimately affects their representatives of government, dynamic for extra ecofriendly fortification. The legislation of government takes decision for the development of protecting the natural environments, which is one of the greatest real behaviors of protecting biodiversity [13].

7.11 Role of science and technology

The role of science and technology is indispensable in conserving biodiversity. As the society develops, it continually uses more resources, which sustains natural biodiversity, but development also leads to improved science and technology. Currently, science and technology are two utmost vital gears in conservation biology. The utilization of science and knowledge, precisely ecology, is important to recognize the relations in the biomass. By considering, these scientists-connections are bright to locate the vital-species in bio-ecosystems. These data are utilized to control protection vitalities. They are correspondingly utilized to know contamination and their dropping properties in the interior of an ecosystem. The biomagnification of pollutants in a food chain is the source of enormous complications for topmost predators. That is an always acclimatizing arena of knowledge, and these two illustrations are just ways to utilize the information to expos. Bioinformatic and technology are fetching extraordinary imperative in bio-conservation. Powerful scientific technologies, like biodegradable packaging, recycling, or renewable energies, help decrease the influence on the atmosphere. In addition, scientific technologies like duplication and replication give experts the capacity to transport again the species that are previously considered vanished. However, biodiversity in natural ecosystems is of the greatest importance. It furnishes and delivers the capitals and amenities that are trusted on each new day. The expansion and suburbanization of persons carry a selfless hazard for natural biodiversity. If no steps are taken for minimizing these variations, there will be catastrophic outlays. There are countless works that can be done by science, politics, and also in the daily lives can provide aids and solutions to these problems. As per humans-being wants to comprehend the jeopardies related by individuals' uncontrollable existence and effort tough to solve, what is previously spoiled and avoid forthcoming damage. The time has come for all of us to bond and preserve biodiversity [14].

7.12 Modern sensors by combining machines

Inexpensive and accessible sensors are accelerating data acquisition in animal ecology. These technologies hold great potential for large-scale ecological understanding. It is argued that zoologists and bio-ecologists can take advantage of bulky datasets created by current devices by the combination of mechanism and learning tactics by field information. Combining appliances and learning into environmental systems could enhance involvements for ecological and zoological models and supervise united hybrid modeling apparatuses. This method will involve nearby inter-penalizing teamwork to certify the excellence of original styles and provide training to the novel generation of data scientists in bio-ecology and protection and management [15].

7.13 Camera traps

7.14 Traditional methods

For pursuing biodiversity, different traditional procedures have been used, such as photographic camera tricks, in which a numerical photographic camera is attached to an infrared sensor to take pictures and videotapes of moving fauna passing the sensor, or in flying, investigations can be work concentrated and expensive. The mechanics and numeric technologies emphasized by the investigations could help decrease the time duration and income obligatory to perceive flora and fauna, although augmenting the efficiency of preservation labours [16].

7.15 Artificial intelligence (AI)

It is progressively utilized to analyze great quantities of protection statistics, such as photographic camera traps, satellite teleimages and drone imageries or audio and video footages, which greatly expand wildlife documentation and intensive care. The non-profit software 'Wild-Me' formed a mist or fog-built platform 'Wild-book', which utilizes computer vision and deep learning algorithms to take lots of crowd-source wildlife pictures to recognize types of fauna and separate animals constructed on their exclusive designs, together with streaks, points, buttons or additional significant corporeal structures like scratches or wounds. Photographs are collected in the fog-built platform by experts and additional assistants or are obtained from societal mass media, and after passing the time periods, the material-related individual species will be cultivated as additional people, scientists, researchers and contributors form the image directory. The accumulated facts benefit to notify management activities, although the community can track their preferred or desired animals in the fog-built platform (**Figure 3**) [17].

Figure 3.
Scene labelling applied to a Northern Kenya landscape (Photo credit: Wild Me) [17].

7.16 Use of artificial intelligence (AI) to combat wildlife trafficking

Artificial intelligence (AI) can correspondingly support improvement unwanted or stealing activities. The programming software PAWS (Protection Assistant for Wildlife Security) can take in previous poaching or thieving histories and the topographical figures of the endangered zone to forecast robbers' forthcoming performance, and project plundering menace atlases and best perambulation directions for guardians. All through the first calendar month of its arena tests in Cambodia in the Srepok Wildlife Sanctuary (SWS), the part recognized as utmost appropriate for the re-establishment of leopard, *Panthera pardus orientalis* (Schlegel, 1857) in the Southeast-Asia, PAWS has facilitated guardians double the figure of traps noticed and detached throughout their rounds. Strategies for the upcoming comprise concerning the computer software to inaccessible detection utensils, such as the drones or satellites, to decrease the requirement for human beings to cross the threshold the statistics and increase the opportunity of PAWS to envisage further methods of conservational corruption as well as unlawful cataloging or angling [18].

7.17 Wildbook

The Wildbook was on-going off to recover the chasing of whale sharks, *Rhincodon typus* (Smith, 1828), which was formerly completed by connecting flexible labels to the fauna that it had not never ever appeared on the surface of sea. The display place has since grown up interested in a massive record of innumerable dissimilar species, together with sea green turtles, *Chelonia mydas* (Linnaeus, 1758); manta rays, *Cephalopterus manta* (Bancroft, 1829); oceanic whitetip shark, *Carcharhinus longimanus* (Poey, 1861); blue whales, *Balaenoptera musculus* (Linnaeus, 1758); Indus river dolphins, *Platanista minor* (Owen, 1853); big cats, *Panthera pardus* (Linnaeus, 1758); giraffes, *Giraffa camelopardalis* Brisson, 1762; and zebras, *Equus zebra* Linnaeus, 1758 [17].

7.18 BearID project

The BearID Project (BIDP) is developing a facial recognition software that can be applied to camera trap imagery to identify and monitor brown bears, *Ursus arctos* Linnaeus, 1758, and updated consequent protection actions has been taken. This stands particularly imperative as camera traps are presently incapable to reliably identify specific bears due to the absence of exceptional usual patterns for many species. Therefore, the team members of zoologists and software engineers have established an AI system using particular pictures of U. arctos from British Columbia, Canada and Katmai National Park, Alaska, USA, which intelligently identified 132 distinct bears with an accuracy of 84%. The project be located already working with indigenous-nations in Canada to implement the new tool within bear research and monitoring programs. The ultimate goal is to expand the scope of the facial recognition software to eventually apply to other threatened species (**Figure 4**) [18].

7.19 Use of environmental DNA (eDNA) for biodiversity

The eDNA, temporarily, allows preservationists to gather biodiversity statistics by utilizing DNA from samples of environment, like H_2O, earth dirt, ice or also air. Altogether vital creatures put off traces of their DNA in their atmospheres over their

Figure 4.
The BearID Project (BIDP) is developing a facial recognition software (Photo credit: Melanie Clapham) [18].

feces, skin or hair, together with many others things. A solitary sample may be containing the genetic code of tens or even hundreds of animal species and can be furnished a thorough photograph of a whole ecosystem. Up-to-date learning has exposed that eDNA can be offered an additional well-organized and cost-effective method for the extensive supervising of global environmental biodiversity. In this case, eDNA specimen noticed 25% additional globally mammal species as linked to camera traps, with partial expenses. Similarly, eDNA can inspect the influence of environmental alteration, distinguish imperceptible intimidations, such as viruses or bacteria, and evaluate the complete strength of a bio-ecosystem, which can be utilized for better safety for the environment. The Nature Metrics team, for example, the Lebanon Reforestation Initiative (LRI) team has joined to use eDNA to measure freshwater biodiversity. Ecosystems of lakes, rivers, ponds, and streams receive critical information about an ancient alternated area to update restoration and rehabilitation efforts [19].

7.20 Connectivity for better conservation outcomes

The Arribada and the FieldKit are the ingenuities' goals to utilize the technology reachable by evolving small-price, exposed foundation instrumental systems. Whereas, the Smart Parks and Sensing Clues (SPSC) emphasis by means of networked-sensors to enhance endangered-part by supervising and administration. All national parks do not have elementary internet or cellphone services as national telecommunication networks do not characteristically spread to these threatened zones. To deliver little-porosity, extended-series connectivity, Smart Parks (SP) organizes a great variety of sensors, comprising gate sensors, alarm systems and animals, automobiles and public trackers, which can track unconventionally on solar porosity, spend small amount of energy and are linked to a protected cloistered net-system located in the national parks. The sensor network pathways and widespread variety of statistics and information are gifted to perceive humanoid intromissions. They can sustenance anti-poaching exertions or animal escapes from the endangered range into the public. They can support pre-empt conflicts of wildlife and human. The data are made available in or near real time in a b-application and can help inform operational decisions related to park management. Smart-Parks-Technology (SPT) has been deployed in protected areas around the world and has helped contribute to the conservation of many endangered species, including African forest elephant, *Elephas africana* Blumenbach, 1797; Javan rhinoceros, *Rhinoceros sondaicus* Desmarest, 1822; and Sumatran orangutan, *Pongo pygmaeus* Lacépède, 1799 [19].

7.21 Protection of wildlife gaming

Even though it was not enclosed by the WILD LABS investigation, games and sports can also assist as treasured devices to trigger spectators by life-threatening preservation problems, particularly between a fresher and a more tech-savvy peer group. The Internet of Elephants, for instance, advances a variety of sportif and numerical involvements founded on technical information to involve public who may not have, then, attracted attention in the conservation of wildlife. Its products include Wild Everse (WE), an increased authenticity transportable sport, where companies drive on preservation assignments in the forest and study, how to make apes protected, and Unseen Empire, which has twisted one of the main photographical camera trap lessons into a gaming knowledge. Players-evaluation actual-photographic-camera trick metaphors to recognize various wildlife species, and in the course of study extra around the shocking impression of deforestation, poaching and other anthropological-activities on threatened-iota, as well as the indefinable apprehensive P. pardus [20].

Author details

Farzana Khan Perveen[1]* and Anzela Khan[2]

1 Classes et événement en Sciences (C.E.S.), Reims, France

2 University of Wollongong (UOW), KDU, Malaysia

*Address all correspondence to: farzana_san@hotmail.com

References

[1] Arroyo B, Mateo R, García JT. Trends in wildlife research: A bibliometric approach. In: Mateo R, Arroyo B, Garcia J, editors. Current Trends in Wildlife Research. Vol. 1. Cham: Springer; 2016. pp. 2-10

[2] Eid Al-Adha. Muslims observe the feast of sacrifice. Huffington Post. 2014; 20142–7. (Retrieved: 3 March 2022).

[3] Haberman C. One or more of the preceding sentences incorporates text from a publication now in the public domain. Charles H. (ed.) Animals in Christian Art. Catholic Encyclopedia. New York: Robert Appleton Company. 1913;12–24 [Retrieved 10 March 2022]

[4] Scanlon J. The world needs wildlife tourism, but that will not work without wildlife. The Guardian. 2017:3-17

[5] Sustainable T. How tourism benefits nature and wildlife-sustainable travel International. 2021;01–25. Online: sustainabletravel.org. Archived from the original

[6] Ceballos G, Ehrlich PR, Raven PH. Vertebrates on the brink as indicators of biological annihilation and the sixth mass extinction. Proceeding of National Academy of Science USA. 2020;**117**: 13596-13,602

[7] Online. https://www.iucnredlist.org/en. 2022 [Retrieved 8 September 2022]

[8] Pacheco X. How technology can transform wildlife conservation. In: Pacheco M, editor. Green Technologies to Improve the Environment on Earth. London, UK: IntechOpen, Rijeka, Croatia; 2019. pp. 34-67. DOI: 10.5772/intechopen.8235

[9] Greenfield P. Humans exploiting and destroying nature on unprecedented scale: report. The Guardian. 2020; 20204–11 [Retrieved 10 September 2022]

[10] Available from: https://www.wwf.org.uk/learn/wildlife/endangered-animals. 2022 [Retrieved 8 September 2022]

[11] Perveen F, Karimullah A, Anuar S. Long-tailed macaques, Macaca fascicularis (Primate: Cercopithecidae): human-monkey behavioral interaction in botanical gardens Penang, Malaysia. Annals of Experimental Biology. 2014; **2**(1):36-44. Available from: http://aexpbio.com/vol2-iss1.html

[12] Perveen F, Abid M. Asian black bear, Ursus thibetanus: Human-bear conflict in the Palas Valley, Kohistan, Pakistan. International Journal of Farming and Allied Sciences. 2013;**2**(24):1172-1178 Online: ijfas.com/wp-content/uploads/2013/12/1172-1178.pdf

[13] Perveen KF, Shah HA. Hunting and trapping pressures on the Himalayan Goral, *Naemorhedus goral* (Hardwicke) (Artiodactyl: Bovidae) in Kohistan, Pakistan. American Journal of Zoological Research. 2013;**1**(1):5-11. Online: http://pubs.sciepub.com/ajzr/1/1/2

[14] Available from: https://greentumble.com/10-ways-to-conserve-biodiversity/ 2022 [Retrieved 8 April 2022].

[15] Tuia D, Kellenberger B, Beery S. Perspectives in machine learning for wildlife conservation. . 2022;13,792–822. National Communication. 2022;**13**: 792-822. DOI: 10.1038/s41467-022-27980-y

[16] Perveen F, Khan A. Migratory status of the goral, *Naemorhedus goral* (Hardwicke) (Artiodactyl: Bovidae) in Kohistan, Pakistan. International Journal

of Biodiversity and Conservation. 2013; **5**(10):671-677. DOI: 10.5897/IJBC2013.0595. Available from: http://www.academicjournals.org/journal/IJBC/edition/October_2013

[17] Ahumada JA et al. Wildlife insights: a platform to maximize the potential of camera trap and other passive sensor wildlife data for the planet. Environmental Conservation. 2020;**47**: 1-6

[18] Lahoz MJJ, Magrath MJL. A comprehensive overview of technologies for species and habitat monitoring and conservation. Biosciences. 2020;**b073**: 23-29. Available from: https://academic.oup.com/bioscience/advance-article/doi/10.1093/biosci/biab073/6322306

[19] Kathrin P, Švara V, Jung M. Environmental DNA-based methods in biodiversity monitoring of protected areas: application range, limitations, and needs. Diversity. 2022;**14**:463-469. DOI: 10.3390/d14060463

[20] NEEF. Environmental education group games and activities. 2022;7a: 25–30. Available from: https://www.neefusa.org/education/ environmental-education-group-games-activities [Retrieved 3 September 2022]

Chapter 2

Freshwater Fish Migration: Fisheries Management Strategy Insight

Gatriay Tut Deng and Birtukan Tsegaye Demisse

Abstract

Migration is a deliberate movement taken by animals for survival. It is commonly categorized as spawning, feeding, and refuge seeking migrations. Migration is governed by costs and benefits. Energy production and utilization is one of the greatest challenges of freshwater fish migration. The upstream and long-distance migrants demand more energy. Orientation and navigation mechanisms in fishes have a long history of interest. Different sensory mechanisms for accurate orientation have been suggested, including orientation using sun position, polarized light patterns, and the Earth's geomagnetic field. Fish morphology plays a significant role in assisting freshwater fish's migration. Long-distant migrants have streamlined body structure and longer caudal regions, while short-distance migrants are fusiform making them hard to move long distance against water current. Since fish migration may involve two different aquatic environments, all migrant fishes that cross the interface between freshwater and saline water habitats must therefore undergo physiological changes. Fish migration activities are influenced by abiotic factors including variations in water temperature, water level, and light availability. Human activities significantly affect fish migration. A good understanding of the migratory behavior of fishes is important for effective fisheries management. Fishermen and near-shore communities need to become aware about the nature of fish migration.

Keywords: fisheries management, fishes migration, freshwater, migration, fish

1. Introduction

Migration is the heroic movement of the animal population from one place to another and it may cover some distances [1]. It is not a simple movement, rather it is a special and deliberate journey taken to achieve some survival goals. Animals migrate for many reasons, over long and short distances and in many different ways such as flying, swimming, walking, or drifting [1]. Migration of fishes is observed in migratory species that carryout this life journey for survival and/or to complete their lifecycle. The long distance migration is demonstrated by the movements of salmonid species between different aquatic ecosystems [2]. Whereas juvenile of many fish species experiences a short daily movements within the same water body between the pelagic and littoral zones, for feeding and avoiding predators [3, 4]. Migration occurs principally between feeding and breeding ground and it may have a strong influence on the population dynamics and affect species composition of the area [5, 6].

Migration usually takes various forms depending on its purpose; namely spawning, feeding, and refuge seeking [7]. Fish migration is fundamentally influenced by spawning migration, which is frequently correlated with temperature, a key regulating element, particularly for species that spawn in the spring. When other mature fish are available, adult fish exhibit what is known as homing behavior, returning to their natal streams to reproduce simultaneously. For migratory fish species to have the best chance of reproducing, synchronized migrations are essential [8]. Spring spawning migration is known in some of the common freshwater fishes like Northern Pike (*Esox lucius* Linnaeus, 1758) [9], Roach (*Rutilus rutilus* Linnaeus, 1758) [10], Perch (*Perca fluviatilis* Linnaeus, 1758) [11], Ruffe (*Gymnocephalus cernua* Linnaeus, 1758), Common bream (*Abramis brama* Linnaeus, 1758) and Bleak (*Alburnus alburnus* Linnaeus, 1758) [12]. Temperature is observed to influence the duration and intensity of spawning migration [10, 12]. In the environment with fluctuating temperatures, light plays an important role to induce spawning migration.

Predation is a strong mortality factor for fish [13] and the major driving force of migration especially for smaller species and juveniles. To reduce predation mortality and maximize growth rate, fish migrate to rivers to avoid predators and seek refuges [5, 14]. Such habitats' change may involve a trade-off between predator avoidance and foraging. If the food accessibility is high, then the fish prey may accept the risk of predation or migrate to an area that gives better protection from predators but low access to food [5, 15].

For migrating fish species, it is important for their survival that they should move freely between feeding and spawning areas. Many fish species cannot complete their life cycles in the lake environment and need access to river habitats [12]. Various management gaps such as natural and artificial barriers can hinder fish to migrate between habitats leading to injuries, decline, or even extinction of fish populations [7]. Manmade barriers particularly have led to fragmentation especially of running water systems. In various freshwater ecosystems, different methods were utilized to overcome these problems. Obstructions like fish ladders, fish elevators, and channel bypasses have been built to facilitate the migration of fish such as cyprinids, esocids, and percids [16]. Misunderstanding of fish migration put migratory fish into a threat for their survival. These fish's species faced great pressures from overexploitation at their breeding habitat and barriers construction that restrict them from continuing their lifecycle. Therefore, this paper is aimed to present the nature of migration in fish survival for effective fisheries management.

2. Costs of migration in fish

Some of the greatest challenges of freshwater fish migration are energy production and utilization. These physiological challenges are magnified during upstream migration where some fish's species spend 75 to 82% of their total energy [17]. Moreover, the energy expenditure during fish's migration increases when the amount of water and its velocity increase at the same time [18]. Binder also added that American shad (*Alosa sapidissima* Wilson, 1811) does not repeat spawning due to high energy depletion. The reason maybe because the forms in which the energy is stored in a fish's body lost their physiological capability to spawn for the second time [19].

Based on the behavior of the migrating fish species, the amount of energy to be spent during migration could be inspected. Some species feed during migration to replace the energy spent while others do not do so. The chance for reproduction rate is very rare for those species that do not feed at the time of migration, sometimes even impossible. Both the success of the migrant species and any of their activities extremely rely solely on the amount of energy they stored before migration. Most of the stored energy is used for muscle movement. During fast swimming and long migration, fish's muscles accumulated lactic acid due to oxygen depletion. The conversion of lactate into glycogen does not last and takes at least 24 hours [17].

3. Orientation and navigation during migration

The mechanisms of orientation and navigation in fishes have a long history that date back from its discovery in the 1930s. This is exemplified by salmon that return from the ocean to their natal streams to spawn, the phenomenon is termed 'homing' [17].

Several orientation and navigation mechanisms have been proposed, and several different cues might likely be involved in orientation during migration. Orientation and navigation activities differ greatly in different water ecosystems. In an open water migration, fishes may probably move in any one of 360° directions. Whereas riverine migrants are located within the channel of the river where they use their added cue of moving either with or against the current, a phenomenon termed positive or negative rheotaxis, respectively.

Quinn [20] suggested various sensory mechanisms for accurate orientation, including orientation using sun's location, polarized light pattern, and the Earth's geomagnetic field. Polarized light may provide a good orientation cue for fishes that migrate at sunrise–sunset when the polarized light patterns are strongest [21]. A response to geomagnetic cues has been shown in some species [22] and biogenic magnetite crystals have been found in the olfactory lamellae of rainbow trout (*Oncorhynchus mykiss* Walbaum, 1792) [22] and in the lateral line of Atlantic salmon [23].

Sockeye salmon (*Oncorhynchus nerka*, Walbaum, 1792) from the Fraser River is magnetically imprinted when they migrate to sea and use variations in magnetic field intensity to successfully locate the coastal imprinting site during their return migration, according to a recent analysis of a long-term fisheries data set [24]. The map and compass hypothesis is not without controversy, though. Fish cannot utilize a compass since they do not have an accurate biological clock [25]. Instead, they use their highly evolved sense of smell to orient to their natal stream.

It has long been known that salmonids utilize smell to discriminate between different streams, even though there is still significant controversy about the sensory cues fish employ to migrate in open water bodies [26]. The pheromone theory [27] and the olfactory imprinting hypothesis [26] are the two main hypotheses that have been put forth to explain olfactory cues. According to the imprinting hypothesis, each stream has a distinctive composition of chemical elements that come from the soil and flora that surround and form the stream, and even a site-specific odor that is imprinted on the youngster. The salmonids may employ this "stream bouquet" as olfactory cues during their return migration from the sea. According to the pheromone hypothesis, migrating adults use population-specific pheromones that are given off by young relatives in their natal stream. According to research on petromyzontid lampreys, adults seek spawning streams by following the bile acid pheromones of young lampreys [28]. Adult lampreys, however, do not return to their natal stream.

4. Mechanisms of migration

4.1 Morphological adaptation for migration

The shape of a fish is a fundamental factor associated with hydrodynamic performance and morphological traits such as a streamlined body can greatly reduce drag thereby assisting swimming during migration [29]. Morphological adaptations for migration in fish have been extensively studied in salmonid migrants. Two key characteristics in the smoltification process are a marked increase in body length relative to mass resulting in streamlining of the body and a decrease in the relative size of pectoral fins [30]. These changes in body form are presumably adaptive for migratory performance. Moreover, variations in body morphology may also occur as a result of adaptation to local conditions and several studies have shown the correlation of body morphology with migration distance and hydrological conditions in the natal habitat [31].

Crossin *et al.* [32] compared the morphology of short- and long-distance migratory populations of Fraser River sockeye salmon. The study result revealed that populations that undertake difficult migrations had short, fusiform bodies that are favorable for reducing transportation costs. Similarly, Brook trout (*Salvelinus fontinalis* Mitchill, 1814) populations that undertook longer migrations were more adapted for energy efficient migration with more streamlined bodies and longer caudal regions [33]. Most of the characters and physiology observed in migratory fish species may be an adaptation to increase survival in the alternative habitats and performance during migration.

4.2 Crossing the interface: Osmoregulation

As migration is the movement requiring migrating species to change its original location, fish move between different water ecosystems. Diadromous migrations between hypo-osmotic freshwater and hyper-osmotic seawater place huge demands on osmoregulatory ability of migrating species. To withstand this osmotic difference, all migrants' fish that cross the interface between freshwater and saline water habitats must therefore undergo physiological changes [34]. The ability to adapt and survive in habitats that differ in salinity is common and has been documented in a wide range of migrating fish species, ranging from Atlantic salmon to pike [35]. However, preparatory physiological adaptations that occur prior to transition to marine habitats differentiate salmonids from other species with euryhaline capability [36]. Though these physiological changes involves the integration of multiple regulations, the increase in euryhalinity is a key characteristic and perhaps the most critical element given the modest osmoregulatory ability of the stream-dwelling parr [36]. Major functional changes in important osmoregulatory organs such as the gill, kidney, gut, urinary bladder, and skin took place prior to the transition between the marine and freshwater ecosystems [37]. Lucas and Baras [38, 39] compiled a list of common long-distance migratory freshwater fish species of Africa, Asia, and South America [**Table 1**]. Moreover the pictorial representation of migratory freshwater fish species is provided in **Figure 1**.

S/N	Scientific and common names	Migration types	Aquatic ecosystem resided	Approx. migration distance (km)
1	*Alestes baremoze* (Characin/ Silverside Joannis, 1835)	Potamodromous	Lake Chad tributaries	650
2	*Alestes dentex* (Characin Linnaeus, 1758)	Potamodromous	Lake Chad tributaries	650
	Brachyplatystoma flavicans (Tiger-Antennenwels Castelnau, 1855)	Potamodromous	Amazon River	3500
4	*Brachyplatystoma vaillanti* (Laulao catfish Valenciennes, 1840)	Potamodromous	Amazon River	3500
5	*Brycinus leuciscus* (Günther, 1867)	Potamodromous	Niger River	400
6	*Hydrocynus brevis* (Tiger-Fish Günther, 1864)	Potamodromous	Lake Chad tributary	100+
7	*Labeo senegalensis* (African carp Valenciennes, 1842)	Potamodromous	Lake Chad tributary	250–300
8	*Pangasius krempfi* (Chinese pangasid-catfish Fang & Chaux, 1949)	Potamodromous	Mekong River	700+
9	*Prochilodus lineatus* (Streaked prochilod Valenciennes, 1837)	Potamodromous	Upper Parana River	600–700
10	*Pseudoplatystoma corruscans* (Spotted sorubim Spiz & Agassiz, 1829)	Potamodromous	Paraguay River	400
11	*Pseudoplatystoma fasciatum* (barred catfish Linneaus, 1766)	Potamodromous	Magdalena River	500–700
12	*Salminus brasiliensis* (Dorado Cuvier, 1816)	Potamodromous	Parana River	850

Table 1.
List of long-distance migratory freshwater fish species of tropical region (Modified from [39]).

5. Factors affecting fish's migration

The movement and distribution of fishes in freshwater habitats can be influenced by living and nonliving factors, such as hydrological regime, temperature, food resources, predation, and competition. The migration pattern of fish is not likely the same since the conditions in the habitat are not static. Fish migration patterns in large floodplain rivers are highly intricate. Fish move not just laterally (in and out of river tributaries, floodplain streams, and various flooded areas), but also longitudinally (up and down the main channel). Among many possible causal factors, flooding is known to significantly affect patterns of fish migration [40].

5.1 Environmental factors that influence migration

Environmental factors greatly influence the pattern of migration in fish. The relative importance of each environmental factor may be dependent on the local characteristic of the habitat in which migration is occurring [18]. For instance, temperature plays significant role in the correlation between the length of the day (photoperiod)

Figure 1.
Pictures of important migratory freshwater fish of tropical region (taken from FishBase.org); (a) Alestes baremoze, (b) Alestes dentex, (c) Brachyplatystoma flavicans, (d) Brachyplatystoma vaillanti, (e) Brycinus leuciscus, (f) Hydrocynus brevis, (g) Labeo senegalensis, (h) Pangasius krempfi, (i) Prochilodus lineatus, (j) Pseudoplatystoma corruscans, (k) Pseudoplatystoma fasciatum, (l) Salminus brasiliensis.

and water temperature. Stephen *et al.* [41] investigated other three factors that have a direct effect on the migratory activities of fishes.

5.1.1 Light

Photoperiod is a seasonal factor that triggers a complex series of physiological events that prepare migrating species for migration. Binder *et al.* [17] reported that any change in light intensity is an essential environmental trigger for starting migration. A population's migratory movement is started and synchronized by the seasonal changes in photoperiod, which give programmed information. Lengthy migrations, as those of Pacific salmon and lampreys for spawning, seem to be principally dependent on photoperiod. Although photoperiod influences the migratory behavior of Pacific salmon smolts, there has been no direct investigation of the role of photoperiod in regulating behavioral changes in Atlantic salmon smolts. In comparison to other vertebrate migrations, photoperiod is observed to increase activity and make fish more responsive to other factors that initiate migration. Hence, photoperiod may function to determine the range of dates when the migration may take place.

5.1.2 Water flow

Changes in water levels to higher volume may function as an initiating factor for migration. When the snow is melting in the spring, water levels become higher and water flow increase as well. Increase in water flow may help the migrating fish species to find the entrance to the stream. However, if flows are very high they can instead be a hinder for fish migration [42]. Very high water flows can also lead to displacements of fish eggs and larvae. Roger and Katherine [43] reported that in spawning migration, the greatest egg deposition occurred when the flood tide is slow. High water flow in rivers may stimulate downstream movement in a large number of fish species and the downstream migration of smolts has been linked to increased water flow. This may be because they migrated at a faster speed and were closer to the water surface, using high turbidity in fast flowing water to make them less visible to predators.

5.1.3 Water temperature

Temperature is another environmental factor that can trigger and synchronize migratory activity in fishes. When temperature function as a stimulus, migration can be viewed as a form of behavioral thermoregulation [17]. This occur under two conditions. Firstly, in thermally heterogeneous environments, the temperature may deviate from the range of thermal tolerance for the migrating species. Thus, fish are forced to migrate for search of new thermally optimal habitat. Secondly, the thermal requirements of migrating fish species may change in demand to undertake various survival activities. For instance, the ideal temperature for growth differs from that of reproduction. In this case, the migration to spawning sites is a result of the temperature. But during spawning migrations, temperature is also a well-known synchronization trigger. Most species with relatively short migrations can attest to this. The springtime upstream migration of sea lampreys (*Petromyzon marinus* Linnaeus, 1758) in the Laurentian Great Lakes is a good illustration of this; the migration does not start until stream temperatures are above −10°C. As a result, the migration starts later in springs that are cooler than usual and sooner in springs that are warmer than average. Thermal thresholds for this kind of spawning migrating fishes is common and believed to have evolved in response to the strict thermal requirements of the developing embryos.

Marine water temperature may have a significant effect on the spawning activities of migratory fishes [43]. Piecuch *et al.* [44] reported that the trout migrating from the reservoir to the main tributary started when the water temperature in the reservoir decreased below 8°C and the water temperature in the tributary is in the range between 5 and 7°C.

5.2 Anthropogenic impact on fish's migration

For their sustenance, migratory species depend on an interconnected chain of intact habitats, a requirement which exposed them to human disturbances in both their habitats and also their migratory routes [45]. Migratory fish species need movement between different habitats, clear migratory pathways between these habitats and appropriate migratory cues (which initiate migration or guide direction) to complete their life cycle or locating valuable resources [46]. Anthropogenic activities have a long history of interfering with fish migration. The complex nature of their life history makes migratory species extremely susceptible to human related threats and climate change, with different impacts at different life stages. Generally, anthropogenic threats to freshwater ecosystems include developmental structures (e.g. dams, weirs, road intersections, hydropower plants), habitat degradation (e.g., land use changes, pollution, channelization), and flow modification (e.g., water extraction, flow regulation, flood control), overfishing, and invasive species (predation, competition, disease, gene transfer, etc.) [39].

The most obvious way in which human activities disrupt migration is through the construction of developmental structures such as dams and weirs that acts as barriers by blocking access to the desired habitats leading to increased migration mortality [38]. Dam construction leaves migratory fishes without any choice since they never pass over it [18]. Therefore, the power of nature to select migratory fishes for their survival may be altered [29]. Human activities in upstream areas blocked with barriers can change migratory cues and downstream habitat through flow control [47]. For instance, tidal regulation can limit sea water incursion into estuaries, capable of reducing suitable breeding grounds for Australian bass (*Percalates novemaculeata* Steindachner, 1866) [48]. Upstream fish migration in watersheds containing such dams has been facilitated by the introduction of fish ladders. Fish ladders, however, only permit migration of fish species with particular phenotypic traits that can overcome the very strong currents [49]. Barriers disrupt stream continuity thereby reducing the abundance and quality of suitable stream habitat. Species hindered by barriers from migration may also be exposed to additional predators when they are enforced to assemble together with other predatory organisms within the same habitat. Another problem caused by the dam is genetic isolation.

Fishing activities and habitat degradation also threaten migratory species particularly diadromous species in estuarine, coastal and marine ecosystems [46]. Large-scale commercial fishing activities commonly occur in these habitats causing overfishing that led to a remarkable decline in many migratory fish populations [45]. Although several studies have reported the importance of estuary and coastal habitats to fishery productions [50, 51], they are vulnerable to anthropogenic activities that may affect water and sediment quality [52, 53].

Land-use changes such as urbanization or agricultural activities can also modify or pollute important habitats along streams [54]. Some pollutants can cause the sensory cells that regulate orientation to sustain physical harm. For instance, heavy metals are extremely toxic to the lateral line and olfactory organs. The metabolism can also be affected by contaminants. Numerous toxicants have been demonstrated to impair fishes' ability to swim by reducing oxygen intake and diverting energy from the swimming muscles. Additionally, this elevated metabolic burden may result in early death [17]. Toxicants and other chemical contaminants [17] may mask the odors that some fishes use to identify the home stream. If a disease or any poison occurs in the water bodies that are dammed, the population of many fishes and other organisms will certainly decline [18].

Climate change as a result of human activities is thought to affect the distribution and productivity of migratory fishes species by interrupting growth, survivorship, habitat availability, migration and [47, 55]. For example, reducing freshwater flow due to climate change might decrease the recruitment and growth of barramundi (*Lates calcarifer* Bloch, 1790) [56] and disrupt the link between sea water and freshwater ecosystems for diadromous fish [47, 57]. Moreover, climate change can reduce survival of diadromous fish in the ocean stage [58] thereby influencing recruitment [59] and decreasing population density. In addition, sea level rise cuased by climatic change may alter the nature of estuarine and reduce freshwater habitats suitability for salmonids [60].

The consequence of climate change on migratory fish particularly diadromous fish species is thought to be complicated for the reason that it influence populations in various ways in different life stages. For instance, Piou and Prevost [61] reported that increasing river discharge and decreasing oceanic growth stage of Atlantic salmon (*Salmo salar* Linneaus, 1758) due to climate change surge their risk of local extinction. Moreover, Hilborn [58] revealed that both the change in ocean climate and anthropogenic activities (dams, fisheries and habitat degradation) contributed to the population decline of Chinook salmon (*Oncorhynchus tshawytscha* Walbaum, 1792) and European eel (*Anguilla Anguilla* Linneaus, 1758).

6. Conclusion

Fishes are widely distributed in both freshwater and marine environment. Fishes like any other animals move from one habitat to another to sustain their life. They undertake this journey for various reasons including spawning, feeding, avoiding adverse environmental condition and predators avoidance. Like other living organisms, fish migrate to cope with their survival needs and requirements. As resources and habitats are not always stable, migration in fishes is one mechanism to compromise with unstable situation in particular habitat. Different age groups in fish have different habitat requirements suitable for their survival. As a problem experienced by many migratory fishes, migratory routes are being blocked by various construction facility such as dams and weirs. To this end, provision of fish's ladder and migratory ways becomes mandatory and should be considered when constructing these developmental facilities. Fisheries experts and managers have focused their attention on these migratory route obstructions to gain good results in fisheries management. A good understanding of the migratory behavior of fishes is an important means of controlling overexploitation of fish stock in particular and enhancing effective fisheries management in general. Catches in the feeding area may increase following the reduction of harvest at the spawning area. Moreover, human activities such as dams and weirs construction obstructing migratory behavior in fishes should be compromised by providing ladders and fish ways, which allow migratory fish to change habitat thereby completing their life cycle.

7. Recommendations

- Awareness regarding the nature of fish migration should be expanded especially to the fishermen and the communities inhabiting the areas near the water bodies.

- Developmental construction like dams and weirs should be provided with fish ladders and way to allow fish effective fish migrations with ease.

- Spawning habitats need not to be exploited to avoid premature harvest of young fishes and brood adult females.

Conflicts of interest

The author declares that there are no conflicts of interest regarding the publication of this paper.

Author details

Gatriay Tut Deng[1*] and Birtukan Tsegaye Demisse[2]

1 Zoological Science, Mekdela Amba University, Ethiopia

2 Zoological Science, Injibara University, Ethiopia

*Address all correspondence to: gatriaytut@gmail.com

References

[1] Hugh D, Alistair DV. What is migration? Bioscience. 2007;**57**(2): 113-121

[2] Dodson JJ. Fish migration: An evolutionary perspective. In: Godin JG, editor. Behavioural Ecology of Teleost Fishes. New York, NY: Oxford University Press; 1997. pp. 10-36

[3] Bohl E. Diel pattern of pelagic distribution and feeding in planktivorous fish. Oceologia. 1980;**44**:368-337

[4] Eklov P. Effects of habitat complexity and prey abundance on the spatial and temporal distributions of perch (*Perca fluviatilis*) and pike (*Esox lucius*). Canadian Journal of Fisheries and Aquatic Science. 1997;**54**(7):1520-1531

[5] Bronmark C, Skov C, Brodersen J, Nilsson PA, Hansson LA. Seasonal migration determined by a trade-off between predator avoidance and growth. PLoS One. 2008;**3**(4):1957

[6] Dingle H, Drake A. What is migration? Bioscience. 2007;**57**:113-127

[7] Meixler MS, Bain MB, Walter TW. Predicting barrier passage and habitat suitability for migratory fish species. Ecological Modelling. 2009:2782-2791

[8] Leggett WC. The ecology of fish migration. Annual Review of Ecological Systems. 1977;**8**:285-308

[9] Vehanen T, Hyvarinen P, Johansson K, Laksonen T. Patterns of movement of adult northern pike (*Esox lucius L.*) in a regulated river. Ecology of Freshwater Fish. 2006;**15**:154-160

[10] Vollestad LA, Labee-Lund JH. Reproductive biology of stream-spawning roach (*Rutilus rutilus*). Environmental Biology of Fishes. 1987;**18**:219-227

[11] Berglund I. Spawning migration of the perch (*Perca fluviatilis L.*), in a subarctic Swedish coastal stream. Aquilo Series Zoologica. 1978;**18**:43-48

[12] Hladik H, Kubecka J. Fish migration between a temperate reservoir and its main tributary. Hydrobiologia. 2003;**504**:251-266

[13] Romare P, Hansson LA. A behavioural cascade: Top-predator induced behavioral shifts in planktivorous fish and zooplankton. Limnology and Oceanography. 2003;**48**:1956-1964

[14] Jepsen N, Berg S. The use of winter refuges by roach tagged with miniature radio transmitters. Hydrobiologia. 2002;**483**:167-173

[15] Borcherding J, Bauerfeld M, Hintzen D, Neumann D. Lateral migrations of fishes between floodplain lakes and their drainage channels at the Lower Rhine: Diel and seasonal aspects. Journal of Fish Biology. 2002;**61**:1154-1170

[16] Ovidio M, Philippart JC. The impact of small physical obstacles on upstream movements of six species of fish. Hydrobiologia. 2002;**483**:55-69

[17] Binder TR, Cooke SJ, Hinch SG. The biology of fish migration. In: Farrell AP, editor. Encyclopedia of Fish Physiology. San Diego: Academic Press; 2011. pp. 1921-1927

[18] Obande RA, Dambo A, Adah PM. Migration in fishes: A review, Nigerian. Journal of Fisheries and Aquaculture. 2014;**2**(1):1-5

[19] Tudorache C, Blush R, De Boeck G. Swimming capacity and energetics of migrating morph and three sticklebacks (*Gasterosteus aculeatus*) and their ecological implications. Journal of Fish Biology. 2007;**71**:1448-1456

[20] Quinn TP. Evidence for celestial and magnetic compass orientation in lake migrating sockeye salmon fry. Journal of Comparative Physiology. 1980;**137**:243-248

[21] Quinn TP, Ogden JC. Field evidence of compass orientation in migrating juvenile grunts (Haemulidae). Journal of Experimental Marine Biology and Ecology. 1984;**81**:181-192

[22] Walker MM, Diebel CE, Haugh CV, Pankhurst PM, Montgomery JC, Green CR. Structure and function of the vertebrate magnetic sense. Nature. 1997;**390**:371-376

[23] Moore A, Freake SM, Thomas IM. Magnetic particles in the lateral line of the Atlantic salmon (*Salmo salar L.*). Philosophical Transactions of the Royal Society of London. Series B, Biological sciences. 1990;**329**:11-15

[24] Putman NF, Lohmann KJ, Putman EM, Quinn TP, Klimley AP, Noakes DLG. Evidence for geomagnetic imprinting as a homing mechanism in Pacific salmon. Current Biology. 2013;**23**:312-316

[25] Stabell O. Migration and navigation. In: Bronmark C, Hansson LA, editors. Chemical Ecology in Aquatic Systems. Oxford: Oxford University Press; 2012. pp. 82-95

[26] Hasler AD, Wisby WJ. Discrimination of stream odors by fishes and its relation to parent stream behavior. The American Naturalist. 1951;**85**:223-238

[27] Nordeng H. Is the local orientation of anadromous fishes determined by pheromones? Nature. 1971;**233**:411-413

[28] Fine JM, Vrieze LA, Sorensen PW. Evidence that petromyzontid lampreys employ a common migratory pheromone that is partially comprised of bile acids. Journal of Chemical Ecology. 2004;**30**:2091-2110

[29] Bronmark C, Hulthen K, Nilsson PA, Skov C, Hansson LA, Brodersen J, et al. There and back again: Migration in freshwater fishes. Canadian Journal of Zoology. 2013;**92**(6):467-479

[30] McCormick SD, Hansen LP, Quinn TP, Saunders RL. Movement, migration, and smolting of Atlantic salmon (*Salmo salar*). Canadian Journal of Fisheries and Aquatic Sciences. 1998;**55**(1):77-92

[31] Taylor EB. A review of local adaptation in Salmonidae, with particular reference to Pacific and Atlantic salmon. Aquaculture. 1991;**98**:185-207

[32] Crossin GT, Hinch SG, Farrell AP, Higgs DA, Lotto AG, Oakes JD, et al. Energetics and morphology of sockeye salmon: Effects of upriver migratory distance and elevation. Journal of Fish Biology. 2004;**65**:788-810

[33] Fraser DJ, Bernatchez L. Adaptive migratory divergence among sympatric Brook charr populations. Evolution. 2005;**59**:611-624

[34] Righton D, Aarestrup K, Jellyman D, Sebert P, Thillart GVD, Tsukamoto K. The Anguilla spp. migration problem: 40 million years of evolution and two millennia of speculation. Journal of Fish Biology. 2012;**81**:365-386

[35] Chapman BB, Hulthen K, Brodersen J, Nilsson PA, Skov C, Hansson LA, et al. Partial migration in fishes: Causes and consequences. Journal of Fish Biology. 2012;**81**:456-478

[36] Stefansson SO, Bjornsson BT, Ebbesson LOE, McCormick SD. Smoltification. Enfield, New jersey: Science Publishers, Inc.; 2008

[37] Clarke WC, Hirano T. Osmoregulation. In: Groot C, Clarke WC, Margolis L, editors. Physiological Ecology of Pacific Salmon. Vancouver, BC: UBC Press; 1995. pp. 317-378

[38] Lucas MC, Baras E. Migration of Freshwater Fishes. Oxford: Black Science; 2001. p. 420

[39] Mota M, Rochard E, Antunes C. Status of the Diadromous fish of the Iberian Peninsula: Past, present and trends. Limnetica. 2016;**35**:1-18

[40] Mochek AD, Borisenko ES, Pavlov DS, Chemagin AA, Budaev SV. Factors affecting the distribution of fish during receding flood in lower Irtysh: Effects of water level and diurnal cycle, International. Journal of Limnology. 2015;**51**:89-100

[41] Stephen MD, Lars HP, Thomas QP, Richard SL. Movement, migration, and smolting of Atlantic salmon (*Salmo salar*). Canadian Journal of Fish and Aquatic Science. 1998;**55**(1):77-92

[42] Ylva L. Upstream fish migration in relation to barriers in Fyrisan [MSc. thesis], Uppsala University, Sweden. 2011; p. 23

[43] Roger RA, Katherine TA. Striped bass spawning in a tidal bore river: The Shubenacadie Estuary, Atlantic Canada. Transactions of the American Fisheries Society. 1999;**128**(4):613-624

[44] Piecuch J, Lojkasek B, Lusk S, Marek T. Spawning migration of brown trout (*Salmo trutta*), in the Moravka reservoir. Folia Zoologica. 2007;**56**(2):201-212

[45] Iwamura T, Possingham HP, Chades I, Minton C, Murray NJ, Rogers DI, et al. Migratory connectivity magnifies the consequences of habitat loss from sea-level rise for shorebird populations. Proceedings of the Royal Society B: Biological Sciences. 2013;**280**:1-9

[46] McDowall RM. Particular problems for the conservation of diadromous fish. Aquatic conservation: Marine and Freshwater Ecosystems. 1992;**2**:351-355

[47] Gillanders BM, Elsdon TS, Halliday IA, Jenkins GP, Robins JB, Valesini FJ. Potential effects of climate change on Australian estuaries and fish utilising estuaries: A review. Marine and Freshwater Research. 2011;**62**:1115-1131

[48] Williams RJ, Watford FA. An inventory of impediments to tidal flow in NSW estuarine fish habitats. Wetlands (Australia). 1996;**15**:44-54

[49] Volpato GL, Barreto RE, Marcondes AL, Andra de Moreira PS, De Barros Ferreira MF. Fish ladders select fish traits on migration still a growing problem for natural fish populations. Marine and Freshwater Behaviour and Physiology. 2009;**42**:307-313

[50] Meynecke JO, Lee SY, Duke NC, Warnken J. Relationships between estuarine habitats and coastal fisheries in Queensland, Australia. Bulletin of Marine Science. 2007;**80**:773-793

[51] Meynecke JO, Lee SY, Duke NC. Linking spatial metrics and fish catch reveals the importance of coastal wetland connectivity to inshore fisheries in Queensland, Australia. Biological Conservation. 2008;**141**:981-996

[52] Zampatti BP, Bice CM, Jennings PR. Temporal variability in fish assemblage structure and recruitment in a freshwater-deprived estuary: The

Coorong, Australia. Marine and Freshwater Research. 2010;**61**:1298-1312

[53] Deegan LA, Johnson DS, Warren RS, Peterson B, Fleege JW, Fagherazzi S, et al. Coastal eutrophication as a driver of salt marsh loss. Nature. 2012;**490**:388-392

[54] Ramírez A, Engman A, Rosas K, Perez-Reyes O, Martino-Cardona D. Urban impacts on tropical island streams: Some key aspects influencing ecosystem response. Urban Ecosystem. 2012;**15**:315-325

[55] Gillson J, Suthers I, Scandol J. Effects of flood and drought events on multi-species, multi-method estuarine and coastal fisheries in eastern Australia. Fisheries Management and Ecology. 2012;**19**(2012):54-68

[56] Tanimoto M, Robins JB, O'Neill MF, Halliday IA, Campbell AB. Quantifying the effects of climate change and water abstraction on a population of barramundi (*Lates calcarifer*), a diadromous estuarine finfish. Marine and Freshwater Research. 2012;**63**:715-726

[57] Schlenk D, Lavado R. Impacts of climate change on hypersaline conditions of estuaries and xenobiotic toxicity. Aquatic Toxicology. 2011;**105**:78-82

[58] Hilborn R. Ocean and dam influences on salmon survival. Proceedings of the National Academy of Sciences. 2013;**110**:6618-6619

[59] Kettle AJ, Asbjorn Vollestad L, Wibig J. Where once the eel and the elephant were together: Decline of the European eel because of changing hydrology in southwest Europe and northwest Africa? Fish and Fisheries. 2011;**12**:380-411

[60] Flitcroft R, Burnett K, Christiansen K. A simple model that identifies potential effects of sea-level rise on Estuarine and Estuary-Ecotone habitat locations for Salmonids in Oregon, USA. Environmental Management. 2013;**52**:196-208

[61] Piou C, Prevost E. A demo-genetic individual-based model for Atlantic salmon populations: Model structure, parameterization and sensitivity. Ecological Modelling. 2012;**231**:37-52

Chapter 3

Status of Hippopotamus, *Hippopotamus Amphibius* L., in the River Sanaga of the Centre Region of Cameroon

Abubakar Ali Shidiki, Tessa Medong Rosalie and Donfack Azabjio Ulrich

Abstract

Wildlife protection and management structures around the world have used protective enclaves often referred to as protected areas to limit pressure on biodiversity. No attention has been paid to species living in unprotected areas often known as free areas. The aim of this study is to contribute to the sustainable management of hippopotamus population in the river Sanaga that is facing threats of being extinct and to also provide information on their status to decision makers. The survey method used during this study was a total count that was carried out using several techniques including foot walk and waterway counts. The results revealed that four hippos were cited in two out of the seven villages in the area. A distance of 32 km was covered in the river. The site with the highest number of hippopotamus was in the village of Tsang with three hippos seen. The main threats to the survival of hippos are poaching, fishing, sand mining and crop farming along the river bank. To lessen these threats on the hippopotamus population, sensitization and awareness campaigns are needed. It is recommended that a hippo sanctuary and a hippo friendly club be created in the study area.

Keywords: Cameroon, hippopotamus, river Sanaga, unprotected area, wildlife protection

1. Introduction

Deforestation has increased in recent decades with population growth and agricultural expansion [1, 2]. Habitat destruction and fragmentation constitute a major cause of wildlife extinction with 40% been mammals [3]. Among the potential resources of these natural environments, wildlife has a prominent place [4, 5]. In some regions of Africa and the world, certain socio-cultural groups contribute to the conservation of wildlife species through their religious beliefs [6–8].

Thus, efforts to conserve wildlife have been motivated by concern over the rarefaction or near-extinction of certain large animals, including the hippopotamus [9]. The main threats to this species are poaching for the meat and ivory/tooth trade and habitat

Figure 1.
Three hippos that were cited at Tsang village.

loss. A field survey showed that in the Democratic Republic of Congo, Hippopotamus populations declined by more than 95% during the 8 years of conflict [10–12].

In Cameroon, as in many other African countries, structures in charge of wildlife protection and management have often used protective enclaves to conserve wildlife. No attention has been paid to species living in free areas [13] and Cameroon is one of these countries where the population trend of Hippopotamus is still unknown. Given the existence of a great spatio-temporal variability leading to the decline of large mammals likely to see their conservation status deteriorate in the near future [14], it seems more than necessary given that the various threats to population of hippopotamus in the Sanaga River is very high and the figures on the state of their population in Africa and in Cameroon in particular is more than alarming [14], it is necessary to carry out this study that has provided information on the status of the hippopotamus population in the Sanaga River and the challenges of their conservation in unprotected areas. These animals also help in plant and nutrient distribution in the ecosystem through excretion and breaking seed dormancy for some tree species [15, 16].

Hippopotamus are amphibians with two species that are known to exist. An adult can weigh slightly above 3 ton depending on the sex [17]. Hippos are social animals and live groups of up 20 animals. It can live up to 40 years in normal conditions. It can give birth once every 2 years with a coup that can weigh up to 50 kg at birth [18]. Hippos graze late in the evenings and early mornings and pass the hold day resting inside shallow waters (**Figure 1**).

2. Methodology

The study was carried out in the Ebebda sub-division 80 km from Yaounde. This sub-division is located in the Lekié Division of the Centre region. It is bounded to the North, South, East and West by Bokito, Sa'a and Monatélé sub-divisions respectively [19]. **Figure 2** shows the map of the study area.

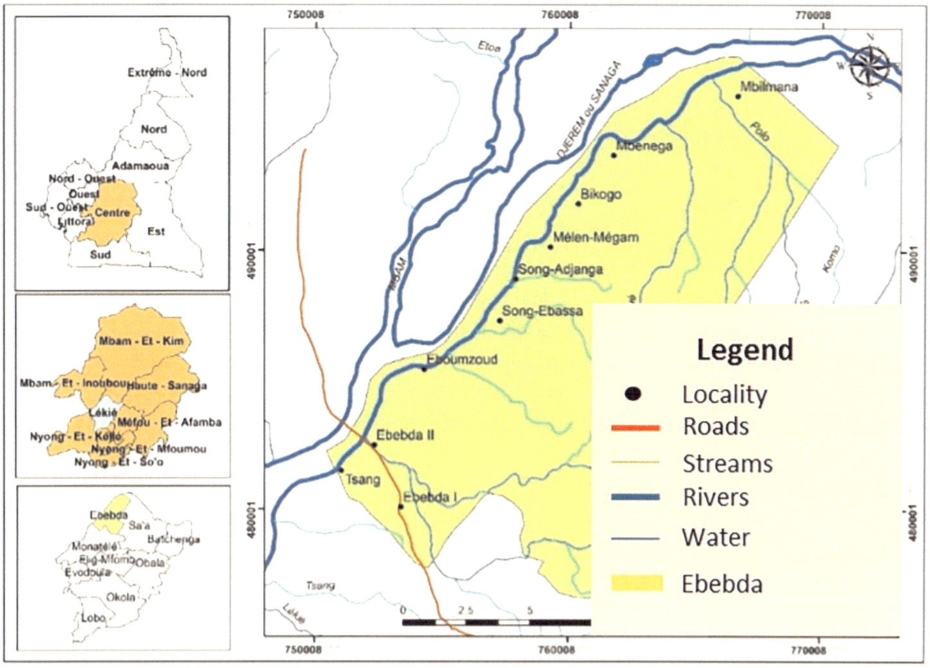

Figure 2.
Map of the study area.

A Guinean equatorial climate prevails in the area with an average annual temperature of 25 ± 2.5°C. The average annual rainfall is 1577 mm [19]. The soils in this area are hydromorphic and ferralitic, characterised by a sandy clay texture. Under forest cover, these soils are sometimes sandy clay, porous, highly permeable, humus-rich and very fertile for crop production. These soils are especially suitable for perennial food crops. Ebebda is watered by the Sanaga River, a fish-filled river (918 km long) with a permanent flow and other small seasonal rivers empting into it such as the river Ngmeh, river Mbe, river Tèt, river Polo, etc. with many streams alike [19]. The area of this basin is estimated to be over 14,000 km^2, with an average flow rate of 2072 m^3/s. It is not really navigable due to many rocks in it.

The vegetation cover is a derived savanna with trees species that are marketable, non-timber forest products (NTFPs) and wildlife species. The flora is very diverse and consists of numerous grasses. The marshy grasslands are mainly colonised by Marantaceae and Zingiberaceae. Wildlife resources in this area are mostly seen in the gallery forests and in the savanna woodlands [19]. The sub-division has 33 villages with an urban centre. The population is estimated at 50,000 inhabitants, including the Eton, Manguissa, Mbamois, Hausa, Bamileke and Fulbe, respectively. The main economic activities of these populations are hunting, fishing, livestock production, trade and sand exploitation.

2.1 Data collection

Based on the following criteria: the presence of hippos in the locality; the accessibility of the villages in relation to the means of travel (canoes); the presence of a sizeable population of hippos, time and financial availability; seven villages were selected for data collection. These were Ebebda I; Ebebda II; Tsang; Mbenega; Ebomzout; Bikogo; Nkolelouga. Data collection was carried out between April and June 2021.

2.1.1 Characterisation of the hippo population

2.1.1.1 Direct observation

This consisted of identifying and assessing the hippos in the different sites along the river banks or Islands. Observations were made early hours of the morning at 5 am to late evenings at 6 pm over a period of 1 week per village along the river site. Canoes were used as a means of transport along the river. At anticipated point along the river with help of a local guide observations were made. The age structure, size of the species and the activities carried out were also recorded.

2.1.1.2 Indirect observation

Hippos can live in the water and on the banks [20, 21]. Indirect observations was based on the walking method, which consisted of following the bank and counting individuals along the river bank [20–22], and the waterway method, which consisted of sailing along the river in a canoe and counting individuals [22, 23], in the five villages visited. The area was divided into two sectors: the Northern sector, which extended from the village of Benga to the village of Mele-megan, which borders Bigoko, where three out of the five islands were visited. The Southern sector which had 11 islands out of which 5 where visited. For the villages Ebeda II and Ebomzoute, they were explored on the shore because they did not have islands. Observations were not made in Ebeda 1 and Nkolelouga because they are urban centres with much human activities. The coordinates of the signs of presence of hippos such as footprints, droppings and tracks were taken into account and marked into a data collection sheet with an accompanying geographical positioning system (GPS) coordinates.

2.2 Socio-economic surveys (baseline survey)

A semi-structured questionnaire was use to interview household heads in the study area. A total of 150 household heads were interviewed in the 5 village of the study area. The criteria for selecting respondents were longevity in the study area, activity (primary occupation), and relationship or links with the hippos. The questionnaire focused on socio-economic characteristic of respondents, knowledge of the hippos, and perception and attitudes towards the species.

2.3 Strategy for managing threats to hippopotamus

The strategy was developed by identifying the actors involved in the conservation of wildlife, categorising them into roles and level of intervention, as well as possible actions taken to protect hippopotamus from human actions.

2.4 Data analysis

The socio-economic data was analysis using SPSS version 16. The cartographic data were entered into QGIS data sheet 2.14.21 and ArcGIS 10.0 for map production.

2.4.1 Determination of hippo populations

2.4.1.1 Relative abundance of hippos

Relative abundance was got by calculating the kilometric index abundance (KAI), also known as the kilometre count index (KCI), which is the ratio between the total number of observations of each sign of activity recorded (*N*) along the sites, and the total distance travelled (*L*) in km.

$$\mathrm{KIA} = \mathrm{Ni}/\mathrm{Lj} \tag{1}$$

Where: KIA = kilometric abundance index for species I; Ni = number of individuals/index of the species along the sites; Lj = length of sites j (in km);
Source: [24, 25]

2.4.1.2 Density

It was calculated using the formula:

$$D = (\text{number of individuals})/(\text{living space area})$$

The surface area of the hippo's living space is given by:

$$S = L * D \tag{2}$$

Where: S = home range area; L = total width in km of the hippo home range estimated at 3 km (i.e. 1.5 km on average on each side of the river); D = the length of the watercourse.

For the specific case of this study, D = 32 km.
Source: [26–28]

2.4.1.3 Propose a management strategy for these threats to hippos in the Sanaga River

In order to develop an effective strategy for the management of hippos in the Sanaga River, a SWOT (Strengths, Weaknesses, Opportunities and Threats) analysis was carried out based on the field results.

3. Results

3.1 Status of Hippopotamus in river Sanaga

The results in **Table 1** below revealed that four hippos were seen in two villages out of the five villages that were surveyed. The village of Tsang had 75% which was followed by Bikogo with 25% of the hippos that were seen. The villages of Ebeda11, Ebomzout and Benga respectively had zero sightings.

Table 1 above showed that the villages Tsang and Bikogo have three and one hippopotamus respectively, as they are located downstream of the river. The absence of these mammals in the villages of Ebebda II, Ebomzout and Benga is due to human activities on the banks of the river and in addition to the infrastructural development.

Village	Number of hippos sighted	Percentage of sighting	Observations
Tsang	3	75	A female, male and young were seen
Ebeda II	0	0	
Bikogo	1	25	A solitary male was seen
Ebomzout	0	0	
Benga	0	0	
Total	4	100	

Table 1.
Number of Hippopotamus observed per village.

Type of index	*D*	*N*	KIA
Footprints	776	32	24.25
Tracks	19	32	0.59
Dung	3	32	0.09
Total	798	32	24.94

N: *number of index;* D: *distance travelled; KIA: kilometric index of abundance.*

Table 2.
KIA of hippopotamus in the study area.

On the other hand, the hippos sometimes sense the presence of humans and do not come out of the water, although signs of their

3.2 Relative abundances

A total of 8 out of 16 Islands in the study area were surveyed, including the river banks of the 2 villages that did not have islands. A distance of 32 km was covered. Kilometric index of abundance (KIA) of hippopotamus which were tracks, droppings and footprints were observed (**Table 2**).

Footprints dominant with a KIA of 24.25 index/km. This is due to the fact that hippos are very huge animals and they can crush nearly every herbaceous vegetation in their path. Tracks with a KIA of 0.59 clues/km are lower because hippos use the same paths in and out of their feeding sites. Dung with a KIA of 0.09 index/km is very low because hippos defecate inside the river which is then carried away.

3.3 Human activities

3.3.1 Agriculture

The majority 80% of the respondents are farmers. The farm lands are not sufficient pushing rural to move into islands inside river which happen to be more fertile; as well as these lands are free not clam by communities. Usually these island are considered private domain of the state (state land). This practice not only destroys the habitat of the hippos but sometimes makes the hippos to become aggressive towards invaders, which often leads to injuries or death on both sides (**Table 3**).

	Abundance of signs of human activity		
Signs of activity	**Number of observations**	**Distance travelled in km**	**KIA**
Battery powered torches	7	32	0.21
Bush fires	24	32	0.75
Campsites	9	32	0.28
Tree cutting	8	32	0.25
Dugout garage	9	32	0.28
Camp fires	21	32	0.66
Gardening	10	32	0.31
Cocoa farming	3	32	0.09
Scarecrows	4	32	0.12
Total	95	32	2.97

Table 3.
Signs of agricultural activities.

These signs indicate the presences of human activities. These signs include battery-operated torches, bush fires, campsites, tree cutting, dugout landing, campfires, gardening and cocoa crops and scarecrows. The overall KIA for agriculture in general is 2.97 index/km. The majority of the signs are distributed between bush fires, campfires and gardening, with KIA of 0.75, 0.66 and 0.31 index/km respectively. The tree felling observed (KIA = 0.25 index/km) is linked to the creation of new agricultural farms. The landing of Canoes and the camps with (KIA = 0.28 index/km) to allow local people to move from one island to another and find shelter if necessary. Scarecrows (KIA of 0.12 index/km) are used to scare animals from farms.

3.3.2 Abundance of signs of poaching activities

The results in **Table 4** revealed the abundance of signs of poaching in the study area.

The results in **Table 4** showed that the local residents are poaching because the number of cartridge shells with KIA of 1.19 index/km showed that hippos are under threats in these localities. Indeed, the hippopotamus is a coveted species in the study area as its meat (bush meat) is a delicacy in the urban centres of Ebada and Yaounde. Respondents said bush meat is a delicacy with hippo carcass that can cost more than a million (1,000,000) CFA francs on the local market.

3.3.3 Sand exploitation

The results in **Table 5** showed the abundance of evidence related to sand mining

The results in **Table 5** showed that sand mining activity is abundant in the study area with KIA of 1.91 index/km. This activity is practiced on a large scale in the study area. Indeed, the sand from the Sanaga is of good quality and is much sought after by real estate developers for construction purposes. Thus, the high demand from sand in the market pushes miners to go to the Islands to get it because at this level, it is still of

Sign of poaching	Number of sightings	Distance travelled (km)	KIA
Cartridge shell	38	32	1.19

Table 4.
Abundance of signs of poaching.

Sand mining	Number of observations	Distance travelled in km	KIA
Sand depot	38	32	1.19
Canoes	23	32	0.72
Total	61	32	1.91

Table 5.
Abundance of evidence related to sand mining.

Fishing	Number of observations	Distance travelled in km	KIA
Fishing nets	12	32	0.37
Fishing canoes	26	32	0.81
Total	38	32	1.19

Table 6.
Signs of fishing activity.

better quality. This activity has led to the degradation and fragmentation of hippo habitats within the Islands.

3.3.4 Fishing

The results in **Table 6** shows fishing activities in the study area with a KIA of 1.19 index/km. In fact, fishing here is artisanal. None of the local residents have a fishing licence, although this is requirement by the Ministry of Livestock, Fisheries and Animal Industry (MINEPIA). According to the respondents fishing is practised early in the morning and that disturbs the hippopotamus that are returning from their pastures and this has often leads to conflicts.

3.3.5 Abundance of infrastructural activities

The construction of houses and related infrastructure like the bridge over the river Sanaga are the some of the activities seen in the study area. The results in **Table 7** show the construction of houses around the river bank with an KIA of 0.78 index/km is the activity responsible for the fragmentation of habitats and the displacement of hippos in the Sanaga river.

3.3.6 Relative abundance of anthropogenic activities

The results in **Table 8** summaries the anthropogenic activities identified in the study area.

The results in **Table 8** revealed that agriculture is the most abundant anthropogenic activity (KIA = 2.97) followed by sand mining, fishing and poaching respectively.

Activity	Number of observations	Distance travelled in km	KIA
Houses	24	32	0.75
Bridge	1	32	0.03
Total	25	32	0.78

Table 7.
Shows the infrastructural activities.

Activity	Number	Distance (km)	KIA
Agricultural activities	96	32	2.97
Poaching	38	32	1.19
Sand pits	61	32	1.91
Fishing	38	32	1.19
others	25	32	0.78
Total	258	32	8.06

Table 8.
Abundance of anthropogenic activities.

3.3.7 Density

Density here is equal to the living area.

$S = 3 \times 32$ km S = living area (km^2)

$S = 96$ km hence

$D = 0.041$ individuals/ km^2

Relationship between anthropogenic activities and hippopotamus Index **Figure 3** shows the relationship between these two elements.

The above graph shows that there are hippopotamus Index within human settlements. The relationship between these two parameters is a straight line on the equation $y = -0.0206x + 1.412$ whose correlation coefficient $R^2 = 0.0014$. The low correlation observed (R^2) echoes that human activities have a high influence on hippopotamus's presence in the study area. Indeed, direct observations in the

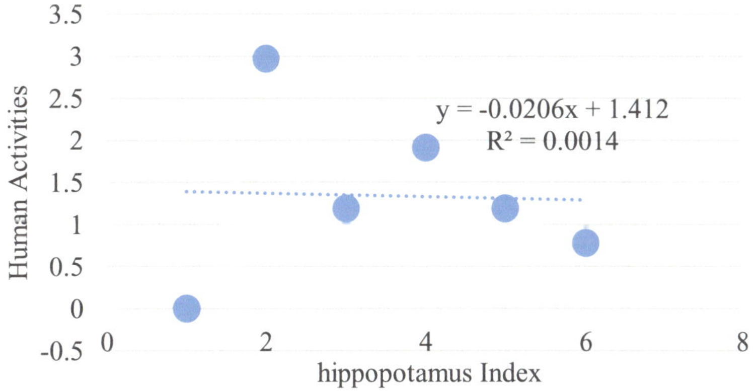

Figure 3.
Correlation curve between human activities and hippo presence.

field revealed that economic activities at community level are on the increase. However, the local populations have also complaint on the destruction of their crops by hippos.

3.4 The presence of hippopotamus

The distribution of hippo and their presence index in space was done with the values of the different Index observed and are summarised in the graph in **Figure 3**. This graph showed that there are hippopotamus Index within human settlements. The relationship between these two parameters is a straight line with an equation $y = -0.0206x + 1.412$ whose correlation coefficient $R^2 = 0.0014$. The low correlation observed (R^2) reflects on the fact that human activities have a high influence on hippopotamus's presence. Indeed, direct observations in the field revealed that activities at community level are going well.

3.4.1 Distribution of hippopotamus signs

The distribution of hippo presence Index in space was done with the values of the different Index observed in the field and are summarised in the graph in **Figure 4**.

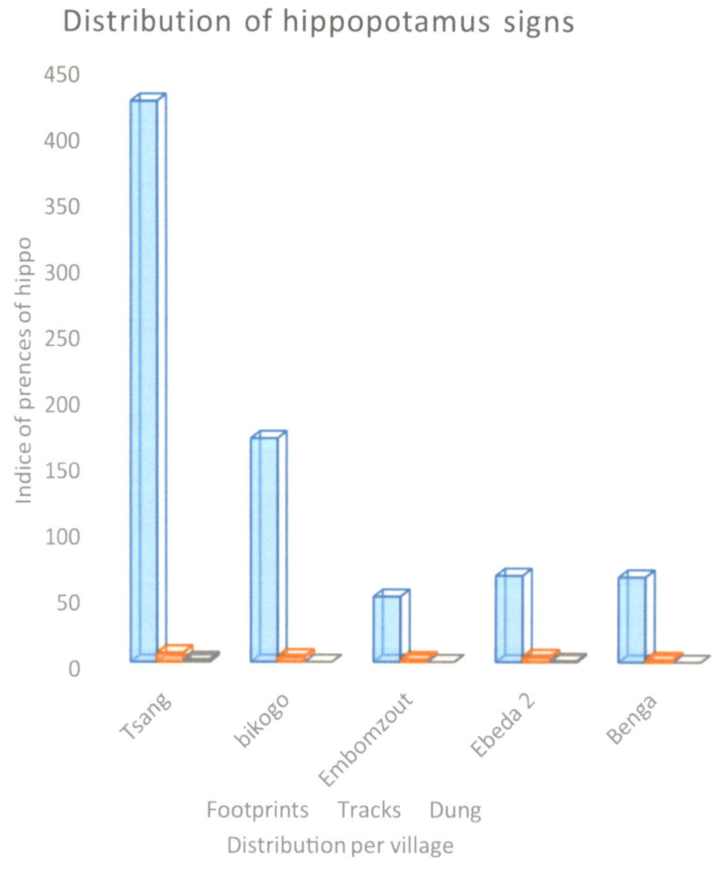

Figure 4.
Distribution of hippo evidence by village.

Figure 4 showed that Tsang village had the majority of hippo evidence. This is due to its geographical position (downstream) of the locality where the water is calmer and agricultural activity is not intense. The village of Bikogo has considerable evidence of hippopotamus, while the villages of Ebomzout, Ebeba and Benga have lesser numbers.

It should also be noted that the presence of droppings here is very low according to the inventories because the hippos may defecate directly in the water which is them carried downstream. Those that are made on the shore are directly collected by the locals for medicinal purposes. **Figure 4** illustrates the layout of the islands along the river and the part of the shoreline where the survey was carried out.

Figure 5 shows the construction of houses and the various sand deposits along the river that putting pressure on the hippopotamus habitat in the study area. The main source of conflicts between humans and hippos along the Sanaga River is that of competition for resources and space.

It should be noted that three out of eight sites were mapped because of the similarity of the problems identified. As a result, agriculture on the islands (hippo habitat) is the most common activity followed by sand mining and artisanal fishing.

3.4.2 Identification of threats to hippos

Despite being on the IUCN Red List of Threatened Species, the hippo is still hunted for its meat and ivory. As a result, it should be noted that habitat loss and fragmentation (45%), human-wildlife conflict (26%), poaching (17%) and fishing (12%) are the main threats to the survival of hippos in the Sanaga River.

3.4.3 Period of hippo's observation

The watching of hippo movements was done using a canoe, which was complicated by the rocky nature of the river bed in the study area, but the field surveys identified footprints of hippos on the river banks.

Figure 6 showed that majority of residents is engaged in farming, fishing and sand mining as their primary economic activity. Majority of these respondents have sighted hippos late evenings (6–8 pm) as seen in the **Figure** 7, which is also the time when hippos are returning from the pastures. During the rest of the day, sightings are less or almost non-existent and only start again in the evenings, i.e. 16:00–18:00, when the hippos start to search of pasture. Farmers (7.68%) generally see them in this time slot because it is the time when they return from their fields. Fishermen come with a 0.63% sighting because this is the time when they go fishing, while sandmen have no sightings at this time. In the late evening, from 18:00 to 20:00, farmers always have a high rate of observations, which is 32.64%. This can be explained by the fact that they camp in their plantation to carry out crop surveillance against hippos. The fishermen with 3.99% of observations can be explained by the fact that some of them practice fishing at night. Lastly, sandmen with a low rate of sighting 1.8% because their activity is mostly carried out during the day.

3.4.4 Hippo poaching

The study area is becoming an increasingly dangerous place for hippos, not only in the waterways they inhabit, but also in their feeding grounds. Although the animal has an indifferent neighbourly relationship with the inhabitants of some villages,

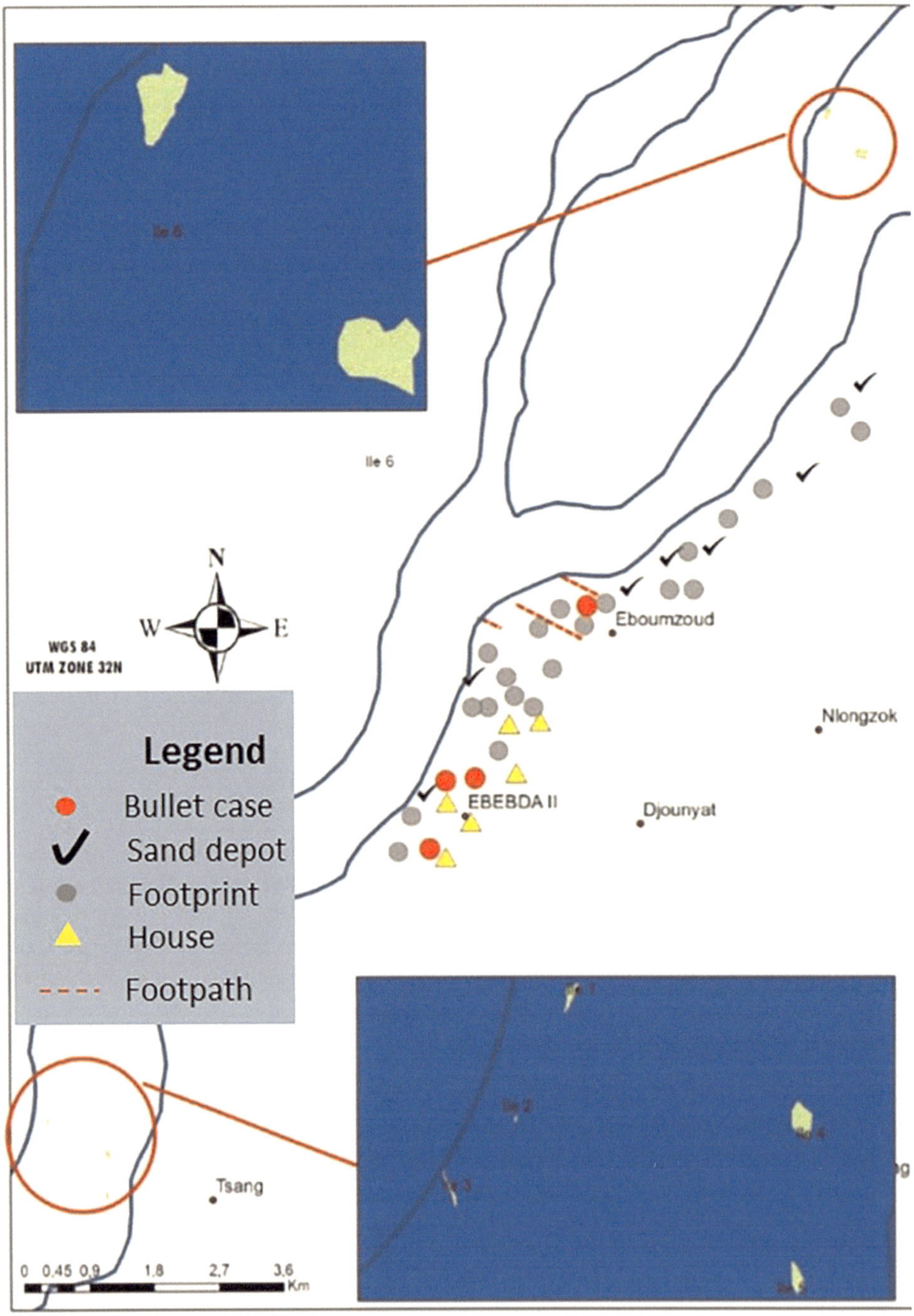

Figure 5.
Map of islands and hippopotamus presences.

especially Ebeda II, it is usually the prey for others who hunt it down for food. In reality, hippopotamus hunting is not organised although there individual hunters who have specialised in the hunting of Hippos in the study area.

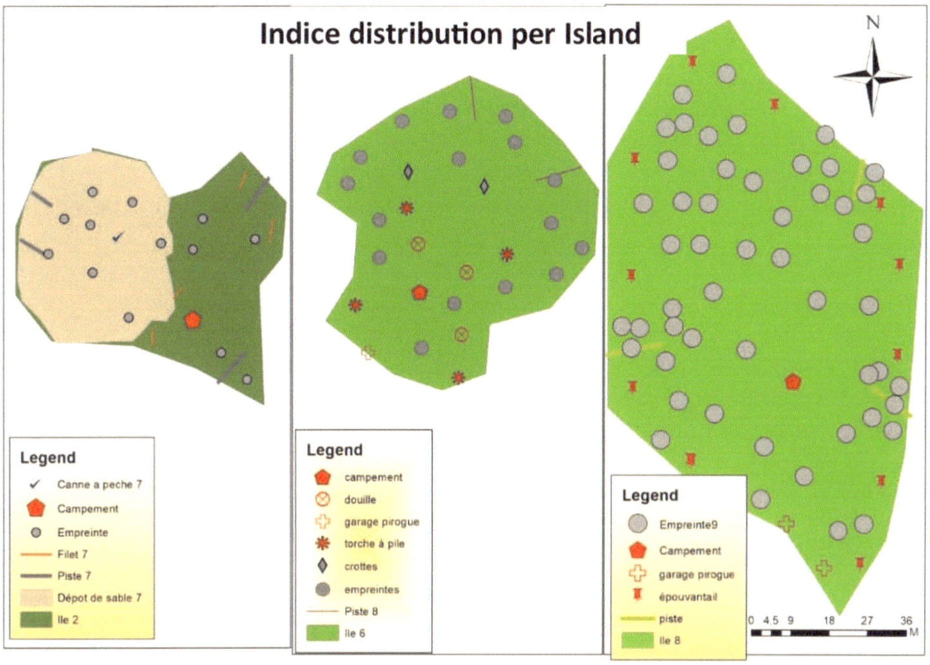

Figure 6.
Period hippopotamus observation.

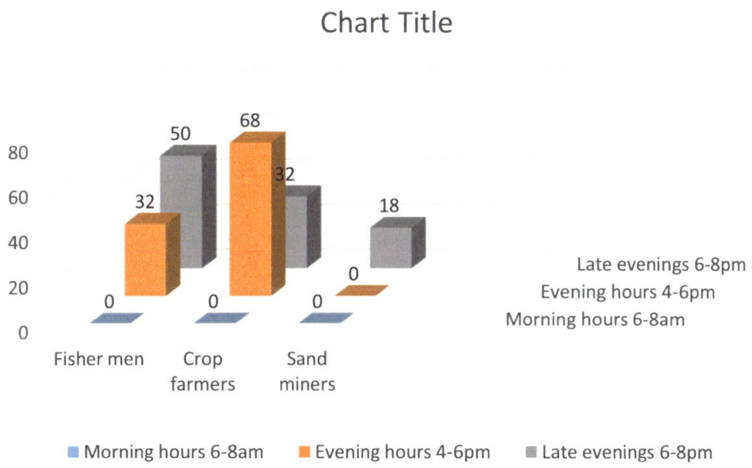

Figure 7.
Map showing human activities and hippopotamus index on the islands of Tsang, Dikogo, Benga villages.

Respondents said large holes are often dug (trap pits) for poaching of hippos, as well as cable traps with large spikes nailed to boards and placed at the edge of hippo grazing sites that has often been used by hippo hunters in the study area. Guns are a favourable asset for some local people who primary occupation is hunting. A hippopotamus was killed in Tsang village in 1997 by the community even though this was reported on the news no arrest was made by the competent authorities.

3.4.5 Agriculture

Agriculture is one of the biggest threats to hippo populations in the study area because local residents are destroying hippo habitats in search of new land for crop cultivation. This may be due to a reduction in rainfall patterns which is observe in the area recently. In addition, the loss of habitat for hippo populations has reduced their rangelands and has caused them to migrate to unfamiliar sites in search of new grazing lands.

3.4.6 Sand mining

Sand mining is also a threat to hippo populations as it has destroyed the habitats through fragmentation. Thus, the sand miners' opinion on the conservation of the animal was negative (**Table 9**).

3.4.7 Human and hippopotamus conflicts

This animals are known to have destroyed vast crop lands and fishing nets etc. this has created conflicts resulting to the killing of this animals. These people are increasingly doubting whether hippos have become so much more important than people and their livelihoods. There is therefore a call for the conservation of hippos in the river Sanaga.

Figure 8 shows that most of the respondents in the villages of Bikogo, Benga, Tsang and Ebomzout (35%) said they have being troubled by hippos during their activities (fishing and gardening) in the sense that they are scared by their size. To this effect, they have called for the killing of hippos by the conservation services. On the other hand, 33.33% of the respondents (Ebebda 2) believe that hippos are not harmful to their economic activity.

3.4.8 Nature of the relationship between people and hippos

The nature of the relationship between humans and hippos can be very beneficial and should allow for sustainable management of the species as hippo dung is used as traditional medicine. Hippos are part of the biological diversity of this area and constitute an asset for man in the regulation of the trophic chain.

Villages	Number of miners per village	Protection of hippos	
		Yes	No
Bikogo	08	0%	100%
Mbega	10	0%	100%
Ebeda II	01	0%	100%
Ebomzoute	05	0%	100%
Tsang	02	0%	100%
Total	26		100%

Table 9.
The perception of sand miners on the conservation of the species.

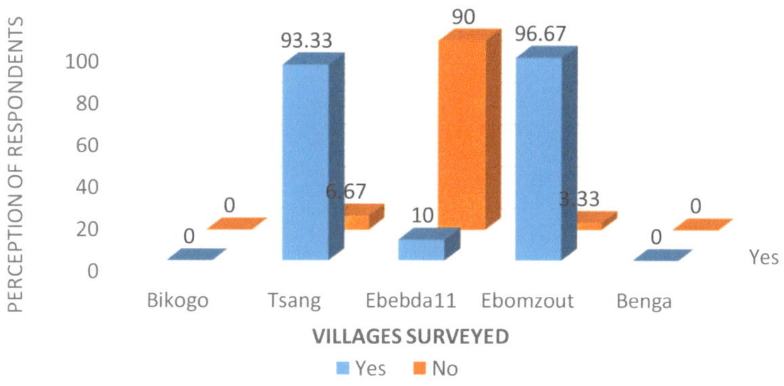

Figure 8.
Residents' perceptions of hippo disturbance.

Strengths	Weaknesses
• Existence of wildlife laws and regulations; • Ratification of international and sub-regional treaties and agreements for the protection of biodiversity; • Existence of hippos in the river;	• The lack of follow-up or total abandonment of the hippos by the services in charge of their protection; • The non-existence of a hippo inventory in the locality; • The lack of data on hippos in the river; • The presence of the area itself not yet classified;
Opportunities	Threats
• The good adaptation capacity of hippos in the river; • The presence of many partners for the conservation of the species; • The creation of a hippo sanctuary in the river; • Easy access to the site	• Presence of agricultural activities, fishing and Sand mining on the hippo habitat; • Poaching practices; • Effect of climate change ; • Loss of its habitat; • Demographic pressure; • Human-wildlife conflict; • Deforestation;

Table 10.
STOW analysis.

3.4.9 SWOT analysis

The Sanaga River has specific bio-ecological, socio-cultural, economic and institutional features that point toward particular considerations (**Table 10**)

4. Discussion

The results in **Table 1** showed that the villages Tsang and Bikogo have three and one hippopotamus respectively, as they are located downstream of the river. The absence of these mammals in the villages of Ebebda II, Ebomzout and Benga is due to human activities on the banks of the river and in addition to the infrastructural development. On the other hand, the hippos sometimes sense the presence of humans and do not come out of the water, although signs of their presence are visible. This result may show an increase compared to the inventories carried out by the Ebogo site manager [29] where one hippopotamus was seen in the Nyong River precisely in Mbalmayo. This discrepancy may be due to the fact that the along the river Sanaga there are more Islands compare to the river Nyong. In addition, the study took place at the beginning of the rainy season, a period during which the Sanaga River high water level is high which makes navigation and counting difficult compared to the study conducted by Amoussou et al. [13] in (3) villages in Southern Benin regrouping groups of hippos isolated in wetlands in the Mono and Couffo Divisions in the dry season where he saw 30 hippopotamus.

The Sanaga River is rich in fish which is a source of animal protein for majority of residents. Fishing is often at night or early in the mornings when hippos are most often seen (**Figure** 7). Some fishermen reported that they have been attacked by hippos at least once during fishing expedition because they fish in shallow waters which are often the resting sites of hippos. The hippos often destroy the nets along their path as well as the Canoes and fish traps. Thus creating tensions between the two parties.

In summary, agriculture is the main threat to the hippo in the study area which is in contrast to the findings as observed by [30] in Benue National Park where poaching is the major threat to hippos as they are been hunted for their meat which is a delicacy and their skin is processed into leather for the manufacture of shoes and bags.

5. Conclusion

The present study makes available a previously non-existence of scientific database on hippopotamus in the Sanaga River with the overall objective of contributing to the sustainable management of hippopotamus in the Sanaga River. The results revealed that, four hippos were seen over the 32 km covered in the river. It also appears that the hippopotamus go out in the evening in search of pasture and do return late at night; and the threats these animals are phasing include agriculture, Sand mining, fishing and poaching.

Anthropogenic activities have shown to have a negative but small influence on the hippopotamus presence. Strategies such as: strengthening the various stakeholders in hippopotamus conservation, enforcing anti-poaching laws, monitoring and protecting hippopotamus and conserving habitats, and enhancing the value of the hippopotamus will help in the mitigation of these threats.

Acknowledgements

We are grateful for the material and logistical support provided by the Laboratory of Environmental Geomatic, Department of Forestry, The Faculty of Agronomy and Agricultural Sciences of the University of Dschang, Cameroon. Special thanks go to all those who participated in the focus group discussions for the vital information they provided that enabled us to move forward with this study.

Conflict of interest

The authors declare that there are no conflicts of interest.

Author details

Abubakar Ali Shidiki*, Tessa Medong Rosalie and Donfack Azabjio Ulrich
Laboratory of Environmental Geomatics, Department of Forestry, The University of Dschang, Dschang, Cameroon

*Address all correspondence to: shidikia@gmail.com

References

[1] Alexandre D-Y. La survie des forêts tropicales. La Recherche. 1992;**23**(244): 693-702

[2] Shidiki AA, Tchamba MN. Small ruminant grazers and stakeholders thoughts in the sustainable management of biodiversity in the kilum montane forest of the northwest region of Cameroon: An overview. In: Kassmi K, editor. Research Developments in Science and Technology. Vol. 6. London UK: PB International; 2022. pp. 124-134. DOI: 10.9734/bpi/rdst/v6/2380B

[3] UICN. La liste rouge de l'UICN des espèces menacées: Un outil fondamental pour la conservation. UICN RED List. 2008

[4] Marchandeau F. Faune sauvage et faune domestique en milieu pastoral une synthese bibliographique (I). Gibier Faune Sauvage. 1992;**9**:67-186

[5] Waitkuwait W. Restauration d'un écosysteme forestier: contribution de l'aménagement de la faune. Compte rendu séminaire sur l'aménagement intégré des forêts denses humides et des zones agricoles périphériques1992. pp. 203-214

[6] Akpona HA, Djagoun CA, Sinsin B. Ecology and ethnozoology of the three-cusped pangolin Manis tricuspis (Mammalia, Pholidota) in the Lama forest reserve, Benin. 2008

[7] Kpetéré J. Valorisation écotouristique des mares aux hippopotames d'Affon: Forêt classée de l'Ouémé Supérieur (Bénin). vol. Master, Ecole de faune, Garoua, Garoua, Cameroun. 2009. pp. 85

[8] Kpéra GN, Aarts N, Tossou RC, Mensah GA, Saïdou A, Kossou DK, et al. 'A pond with crocodiles never dries up': A frame analysis of human–crocodile relationships in agro-pastoral dams in Northern Benin. International Journal of Agricultural Sustainability. 2014;**12**(3): 316-333

[9] Asibey EO. Wildlife as a source of protein in Africa south of the Sahara. Biological Conservation. 1974;**6**(1):32-39

[10] Dibloni TO, Vermeulen C, Guenda W, Alfred MN. Structure démographique et mouvements saisonniers des populations d'hippopotame commun, Hippopotamus amphibius Linné 1758 dans la zone sud soudanienne du Burkina Faso. Tropical Conservation Science. 2010;**3**(2):175-189

[11] Jean-Pierre BM, Shidiki AA, Tchamba MN. Analysis of current Governance in the Sustainable Management of the Virunga National Park of the Democratic Republic of Congo. International Journal of Forest, Animal and Fisheries Research (IJFAF). 2021;**5**(5). ISSN: 2456-8791. DOI: 10.22161/ijfaf.5.5

[12] IUCN. The IUCN Red List of Threatened Species. Version 2014.3. Available from: www.iucnredlist.org. Consulté le 5 mai 2015. 2015

[13] Amoussou K, Mensah G, Sinsin B. Données biologiques, éco-éthologiques et socio-économiques sur les groupes d'hippopotames (*Hippopotamus amphibius*) isolés dans les terroirs villageois en zones humides des départements du Mono et du Couffo au Sud-Bénin. Bulletin de la Recherche Agronomique du Bénin. 2006;**53**:22-35

[14] Iyah NE. Status et tandence collective de la population de hippopoatamus (*Hippotamus amphilus*) au parc national du Faro et sa peripherie

au cameroun. Universite polytechnique de Bobo dioulasso. 2015. p. 27

[15] NODDEL et PNDP. Plan de Développement Communal d'Ebebda PNDP-NODDEL, Cameroun. 2015. pp. 193

[16] Bauer S. Hoye BJ Migratory animal's couple biodiversity and ecosystem functioning worldwide. Science. 2014;**344**

[17] Eltringham SK. Les cochons. Pécaris et hippopotames. Enquête de situation et plan d'action pour la conservation. 1993. pp. 55-60

[18] Kabre ATL, Koné HS, Nandnaba S, Wendyermé P, Bobodo BS. Gestion de la zone d'interface écologique faune-population: le cas de l'Hippopotame au Burkina. Annales de l'Université d'Ouagadougou-Série C. 2006;**04**:177-205

[19] Doughty CE, Roman J, Faurby S, Wolf A, Haque A, Bakker ES, et al. Global nutrient transport in a world of giants. Proceedings of the National Academy of Sciences of the United States of America. 2016;**113**:868-873

[20] Eltringham SK. Review of priorities for conservation; action and future research on hippopotamuses. In: Oliver RL, editor. Pigs, Peccaries and Hippos Status Survey Action Plan. Cambridge: IUCN; 1993. pp. 127-148

[21] Kabre AT, Poda WC, Kone L. et les populations riveraines dans la zone d'interface écologique. Annales: Lettres, sciences humaines et sociales. Série A. 2006;**4**:357

[22] Ngog Njc J. Contribution à l'étudc dc la structure de la population des hippopotames (*Hippopotamus amphibius* L.) au Parc National de la Bénoué (Cameroun). 1988

[23] Bouché P. Inventaire total aérien dans le site de la Réserve de Biosphère de Mare aux Hippopotames. PAGEN/MECV, Burkina Faso. 2005

[24] Attwell GR. Some observations on feeding habits, behaviour and inter-relationships of Northern Rhodesian vultures. Ostrich. 1963;**34**(4):235-247

[25] Buckland ST, Anderson DR, Burnham KP, Laake J. Distance Sampling: Estimating Abundance of Biological Populations. London: Chapman & Hall; 1993

[26] Preatoni DG, Tattoni C, Bisi F, Masseroni E, D'Acunto D, Lunardi S, et al. Open source evaluation of kilometric indexes of abundance. Ecolog ical Informatics. 2012;**7**(1):35-40. DOI: 10.1016/j.ecoinf.2011.07.002

[27] Bouchie P. Méthodes d'inventaire de la grande faune à l'usage des ZCV. Université de liégé; 2008. p. 7

[28] Meli M. Nouvelles Technologies Et Amelioration Des Inventaires Fauniques: Cas Des Drones Et Camera-Pieges Dans La Gestion Durable Des Hippopotames Au Parc National De La Benoue. Dschang: Département de Foresterie. Université de Dschang; 2019. p. 90

[29] MINTOUL. A survey of Hippopotamus in the rive Nyong a potential for eco-tourism. The Ministry of Tourism Cameroon report 2014. 2014

[30] Scholte P, Iyah I. Declining population of the vulnerable common hippopotamus *Hippopotamus amphibius* in Bénoué National Park, Cameroon (1976–2013): Thc importancc of conservation presence. Oryx. 2016; **50**(3):506-513. DOI: 10.2305/IUCN. UK.2017-

Chapter 4

Spatio-Temporal Distribution of the Black Rhino (*Diceros bicornis L.*) in the Midlands Black Rhino Conservancy, Zimbabwe

Blessing M. Mugaviri, Gilbert H. Moyo, Ezra Pedzisai and Cuthbert Maravanyika

Abstract

Geographic Information System (GIS) and Remote Sensing (RS) technologies have many attributes that are beneficial in detecting, mapping, and, monitoring change in Land use/Land cover (LULC). This study used the technology with the aim to evaluate the Spatio -temporal impacts of Land use/Land cover Changes (LULCC) on Black Rhino distribution in Midlands Black Rhino Conservancy (MBRC), Zimbabwe. The study used time series satellite data. Landsat images were downloaded for the month of May at five-year intervals from 2000 to 2020. LULC and Normalized Differences Vegetation Index (NDVI) maps obtained were used in change detection. The images were classified using QGIS software on the maximum likelihood classifier algorithm. Presents and absence data for Black Rhino was used for distribution mapping. Quantum Geographic Information System (QGIS) and, R studio software were used for analysis. Results indicated that, a big percentage cover change was the bare land which increased by over 160%. Woodland decreased by about 46% within the same space of time. LULCC showed a significant positive relationship with black rhino distribution ($p = 0.0381$). MOLUSCE plugin was used for Prediction of LULCC for the year 2030, results indicated the highest increase in bare land 16.59%.

Keywords: biodiversity, habitat, land cover land use changes (LCLUC), Spatio-temporal, Black Rhino

1. Introduction

Land Use and Land Cover, Geographic Information System (GIS) and Remote Sensing (RS) technologies have many attributes that would be beneficial in detecting, mapping, and monitoring change in Land use/Land cover (LULC). This study used the technology with the aim to evaluate the Spatio -temporal impacts of Land use/Land cover Changes (LULCC) on Black Rhino distribution in Midlands Black Rhino Conservancy (MBRC), Zimbabwe. The research gives a better understanding of the ecosystem for sustainable management. Remote Sensing using space-borne sensors is a tool, par excellence, for obtaining synoptic observations on the spectral behavior of various environments, for instance, land surface changes (degradation), water quality, soil and, atmosphere [1].

LULC represents an important factor in environmental analysis and spatial planning approaches [2]. When discussing the environment there is a need to understand the existence and importance of each species in an ecosystem and bear in mind that habitat loss threatens the existence of fauna and flora [3]. LULC is a dynamic variable because it reflects the interaction between socio-economic activities and environmental changes, for example, where deforestation has taken place and where land has been cleared due to anthropogenic factors [1]. For this reason, it is necessary to be updated frequently. Integrated GIS and RS have already successfully been applied to map the distribution of several plant and animal species, their ecosystems, landscapes, bio-climatic conditions, and factors facilitating invasions [1, 3–6]. Remote sensing imagery is available for most parts of the world since 1972. The multidate nature of satellite imagery permits monitoring dynamic features of landscape environments and thus provides a means to detect major land cover changes and quantify the rates of change [7]. However, there are inadequate researches that highlight land use impacts on specific species distribution.

The interpretation and analysis of Landsat TM image since 1987, provide comprehensive information regarding the various land uses and the associated environmental problems [8], for example to determine the land-use changes due to new settlements, deforestation, and erosion due to land clearing activities RS techniques have been successfully applied [9]. Due to Advancement in satellite sensors, their analysis techniques are making remote sensing systems fruitful, realistic, and attractive for use in research and management of natural resources [7].

This research was conducted in Midlands Black Rhino Conservancy (MBRC), which consists of privately-owned bush and farmland to evaluate if LULCC have got an impact on Black Rhino distribution. The conservancy supports cattle grazing and game utilization. However, there has been a change in land use in some areas. The clearing of huge tracts of land for mining led to the displacement of many animals, environmental degradation, and the irrecoverable destruction of animal habitat, for example, Black Rhino home ranges. A visit to one of the disused plants of the mining company at Two Springs, deep in the conservancy area, shows furrows and heaps of dumps from mining activities, with no reclamation efforts having been done contributing to habitat loss.

Currently, there have been some changes in land-use practices. Agriculture used to be subsistence farming only but, it is no longer subsistence since they have extended the cropland for better yield. These alterations in land use led to a change in the composition of vegetation diversity thereby, raising a flag to research if, the changes have affected wildlife distribution in the conservancy. Since the area was set aside for cattle and game utilization when the conservancy was formed, there is a need to assess change in land use to quantify the LULCC percentage. Studying trends enables an understanding of changes in land utilization.

The rate of biodiversity loss in MBRC is a serious cause for concern to the ecosystem. Human-induced LULC changes have contributed to the dilapidation of Black Rhino habitat in MBRC, hence the need for an evaluation on LULC change in the important conservancy. It is also unknown to what extent the activities being practiced have impacted the Black Rhino distribution and, also how it will affect the distribution in the future. This article quantified LULC change from 2000 to 2020 and also analyzed the relationship between LULC change and Black Rhino distribution within MBRC for the years 2000 and 2020. Furthermore, it predicted the extent of LULC change in MBRC by 2030.

A major reason for researching historical LULCC is that by understanding the past, we can better understand future trajectories for managers to make informed decisions on the management of the ecosystem [1]. This can be achieved using GIS and RS, and diversity indices to observe land-use change. There is a significant gap in our understanding of the spatial and temporal ecology of biodiversity and ecosystem goods and services. This research seeks to fill the gap by evaluating the spatial and temporal impacts of LULCC on the Black Rhino distribution and uses a model to predict future changes.

1.1 Black rhino (*Diceros bicornis L.*)

Rhinos are large odd-toed ungulates that fall into the Perissodactyla order and the Rhinocerotidae family. The black rhino is a large gray animal that stands 1.4–1.7 m and weighs between 996 and 1362 kg. The black rhino in **Figure 1** has an upper prehensile lip that they use for browsing which is a predominant physiological difference between the two African rhino species. This lip enables them to browse selectively on a diverse array of woody species across their range [10, 11]. They have poor eyesight, which they compensate for with an acute sense of smell and hearing Black Rhino have two continually growing horns that are variable in shape and size with the front horn normally longer than the rear horn. Rhino horn is made up of keratin and is used for predator defense, a stake in encounters with other rhino, and a tool for pulling down hard to reach branches for feeding. The rhino is sensitive to its home range and it is ideal to study the effects of LULCC to its home range.

Figure 1.
Image showing a black rhino.

2. Materials and methods

This research study is on spatio-temporal impacts evaluation of LULCC on Black Rhino distribution in MBRC Zimbabwe. This chapter presents the methodology of this research study. The next important step in any research process after the study of literature and identifying the research questions is deciding on the most suitable methodology. The research methodology is the overall approach to the design process from the hypothetical foundations to the collection of data and analysis adapted for a study [12]. The methodology is therefore how we discover how to go about a task of finding out what we believe to be true [13]. This chapter presents the study area, research method, research design, LULC data acquisition and Rhino distribution data collection.

2.1 Study area

The Midlands Black Rhino Conservancy Trust (MBRC) is situated in the heart of Zimbabwe, located at latitude 18°58'01" S and longitude 30°11'24″ E. The area consists of 63,000 ha (156,000 acres) of bush and farmland bounded by the Munyati River on the northern boundary and the Sebakwe River on the southern boundary with Lake Sebakwe and its Recreational Park in the middle. Agriculturally, Sebakwe is in the country's Natural Farming Region whose rainfall ranges from 650 to 700 mm per annum [14]. The temperature ranges between 25°C to 28°C. About 70% of the soils are derived from granite, which is loam and light-textured soils. In **Figure 2** Munyati River drains the area to the north and Sebakwe River cuts through the conservancy and drains the study area to the south. Their subsystems which include, among others; Shorai, Nyamaponde, and Zibagwe Rivers also drain the area from South-East to the North-West [15]. Although there is a rich network of rivers, most of them are ephemeral, and annual, animals and cattle rely on pumped water holes.

The ranch has a variety of soil patches including black clays in Vlei, sandy loams, loams, and red soils in uplands particularly the great dyke. The soils are classified as serpentinites and are rich in Chromium and its associated minerals. The topography is generally flat except for the Mazuri- Chinyika boundary which comprises a stretch of the Great Dyke [14]. Wildlife ranching and cattle farming is the predominant land use whilst crop farming and mining are now being practiced. Different woodland types exist in MBRC. These include Miombo, Mopane, Terminalia, Acacia, and mixed species. MBRC has a diverse vertebrate fauna that includes mammals, birds, reptiles, and fish. Mammals include megaherbivores such as Elephants (*Loxodonta africana*) Black rhino (*Diceros bicornis*), Giraffe (*Giraffa camelopardalis*), Wildebeest (*Connochaetes taurinus*), Zebra (*Equus quagga*), Waterbuck (*Kobus ellispiprymus*), Impala (*Aepyceros melampus*) and Eland (*Taurotragus oryx*). The birds include the recent sighting of the ground hornbill, the Gray Lorie, Yellow-billed hornbill to mention only a few [15].

2.2 Research instruments

The instruments that were used were Satellite images, Global Position System, Vehicle, Digital camera, Datasheet, Pen, pencil, ruler, QGIS 3.4, 2.8 and R studio 4.0.2.

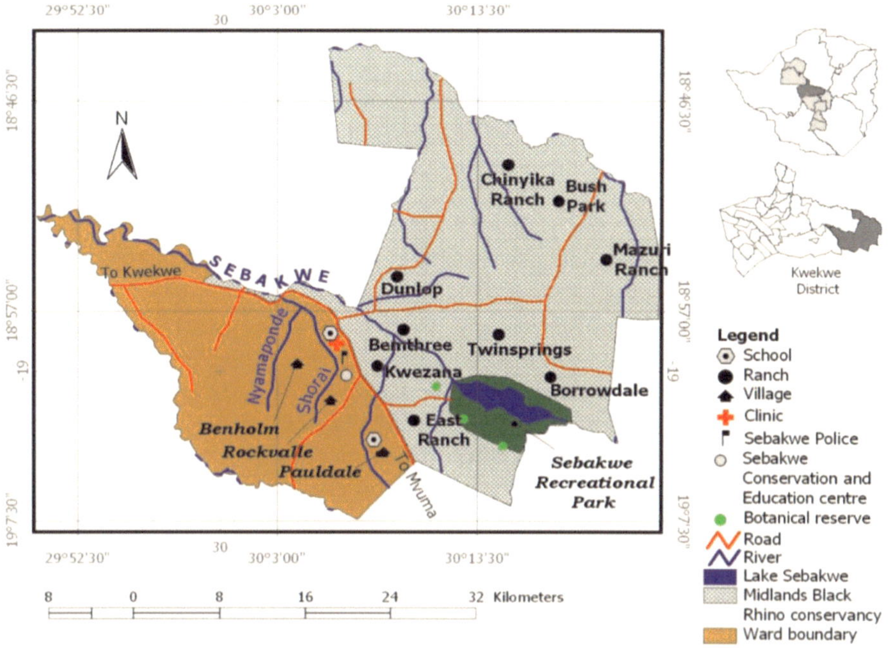

Figure 2.
Map showing midlands black rhino conservancy.

2.3 Research method

In this research study, the researcher adopted a mixed-method research approach. The overall goal of mixed methods research was to combining qualitative and quantitative research components. It also expands and strengthens a study's conclusions and, therefore, contribute to the published literature [13]. In this research study, the use of mixed methods contributed to answering research questions such as the rate of change in LULCC from the year 2000 to 2020 in MBRC, if there is any relationship between LULCC, and Black Rhino distribution within MBRC, and the perceived habitat loss in MBRC by 2030. Qualitative methods were used to explore the phenomena, and generate the conceptual model while quantitative methodologies were used to confirm the validity by testing the hypothesis [16].

2.3.1 Research design

The study included pre-field data collection, field survey and post-field data analysis. Primary data collected during field work and secondary information about the Rhinoceros and the study area were extensively used for the study purpose. To address the research study questions, mixed methods were used, consultations with site farmers, management, and remote sensing were employed as data collection tools. This research study adopted a longitudinal survey approach which refers to an investigation where participant outcomes and possibly treatments or exposures are collected at multiple follow-up times. Longitudinal study generally yields multiple or "repeated" measurements on each subject [17]. Researchers record the information that is present in a population, but they do not manipulate variables [18]. The method is often used to make inferences about possible relationships or to gather preliminary data to support further research and experimentation [19]. Semi-automatic classification plugin in GIS and RS technology were used in detecting, mapping, and investigating the change in LULC. Ground control points were randomly sampled in the study area covering areas with grasslands, cropland, built-up areas, bare land, water bodies, and woodland for ground-truthing [20]. GPS locations for Rhino sightings were collected in the year 2020 and secondary data from year 2000 was used as baseline data.

Landsat images where downloaded from the NASA website (https://earthexplorer.usgs.gov) [21]. Topographic map scale 1:50,000 was used as a guide for interpretation. A digital elevation model of 30 m resolution was used. A statistical sampling methodology based on area frame sampling was adopted for this study. The method relied on satellite imagery, photographs, and maps for data collection [22]. The unaligned area frame sampling scheme was preferred for the area because of the heterogeneous land use cover found in the area. High densification for sample sites where preferred to validate the consistency of the land cover and land use database. Individual land parcels and ground cover classes were identified in each sample segment. The 2030 map was predicted using molusce plugin. Present only and absence data were used to map the Black Rhino distribution.

2.3.2 Rhino distribution data collection

Presence and absence data were collected for spatio-temporal data to plot the distribution of Black Rhino in MBRC on a map. Direct observations of Rhinos and fresh dung were used as present data, furthermore, a handheld Global Processing System (GPS) was used to record the locations of the sightings. Secondary data was also used from the Parks and Wildlife Authority database where sightings and locations are recorded. Secondary data of one hundred, for the year 2000 were used, and 100 sightings for the year 2020 were recorded.

2.3.3 Land use/land cover data acquisition

Satellite images were downloaded from the website (https://earthexplorer.usgs.gov), [21] however, before downloading images to be classified, a pilot survey was conducted to check for images with less cloud cover.It was then observed that for the month of May in each of the following years: 2000, 2005, 2010, 2015, and 2020 the images had less than 10% cloud cover, hence, the images for May were used. The images were georeferenced and fit to the Universal Transverse Mercator (UTM) projection system (zone 45, datum WGS-84). The main steps involved in image classification are determining a suitable classification system, feature extraction, selecting good training samples, image pre-processing and selection of appropriate classification method, post-classification processing, and finally assessing the overall accuracy.

Tiles were obtained in GeoTiff image format for pre-processing. Image composites were generated using images with cloud cover less than 10% from the United States Geological Survey (USGS) archive Landsat–5 Thematic Mapper (TM), Landsat-7 Enhanced Thematic Mapper Plus (ETM+), and Landsat-8 Operational Land Imager (OLI) covering the MBRC area at a spatial resolution of 30 meters, Path 170 and Row 73. Using a shapefile for MBRC the area was clipped from the satellite imagery using Quantum Geographic Information System (QGIS) software version 3.4. Satellite imagery were also pre-processed for radiometric errors. A combination of bands 2, 3, 4, 5, 6, and 7 (**Table 1**) were merged to form composite images for each period under study. Vegetation appears in shades of dark and light green, hot surfaces such as built-up areas and bare land appear in shades of red or yellow. The eligible ranges for candidate images were from 2000 to 2020. Eighty Ground Control Points were collected and a shapefile was created to superimpose on the clipped satellite imagery as reference points for classification.

This research study utilized a supervised classification interpretation approach. According to the research purpose and type of vegetation in the area, six classes including built-up area, cropland, grasslands, bare land, woodland, and water bodies were identified and classified. In this type of classification spectral classes were grouped first, based on the numerical information of the data, and were then matched [23]. Five images were ultimately used to create the image composites for the periods: 2000, 2005, 2010, 2015, and 2020. These images comprised of varying wavelengths separated into wavelength bands (**Table 1**). Band 1 is known as blue as it provides increased penetration of water bodies and also capable of differentiating soil and rock surfaces from vegetation. Band 2 covers the green reflectance peak from leaf surfaces, it separated vegetation (forest, croplands with standing crops) from the soil. In this band barren lands have appeared as brighter (lighter) tones, but forest, vegetation, bare croplands, croplands with standing crops have appeared as dark (black) tones. The third band highlights barren lands. Bands 4 to 7 function in the best spectral regions to distinguish vegetation varieties and conditions in the preceding bands.

Using the QGIS 3.4 maximum likelihood classification, the 4 images were analyzed and processed. Using the raster images, which have values attached to each pixel, training data was created using known sites (Ground Control Points) in the study area, from which the software was able to identify sites with similar cell values. The Maximum Likelihood classification was able to assign each cell in the input raster to the class that it has the highest probability of belonging to, resulting in the creation of 6 land classes: Woodland, mined area, Bare land, Cropland, Built-up Areas, and Water Bodies.

Bands	Wavelength (micrometers)	Resolution (meters)
Band 1-Blue	0.45–0.52	30
Band 2-Green	0.52–0.60	30
Band 3-Red	0.63–0.69	30
Band 4-Near Infrared (NIR)	0.77–0.90	30
Band 5-Shortwave Infrared (SWIR) 1	1.55–1.75	30
Band 6-Thermal	10.40–12.50	60* (30)
Band 7-Shortwave Infrared (SWIR) 2	2.09–2.35	30
Band 8-Panchromatic	.52–.90	15

Source: (Shafri, 2015).
Notes: Colors blue, green, and red indicate the ideal environmental conditions. The color blue indicates water, green is for vegetation, red indicates high reflection (heat), for example from surfaces that are impervious like roads, buildings, etc., bands 4–8 are invisible but affect the intensity of wavelengths in bands 1–3, which then results in varied color shades.

Table 1.
Bands of varying wavelengths separated into wavelength bands as referenced in the study.

2.4 Accuracy assessment

The accuracy of the quantified land cover changes were assessed with the help of reference datasets based on the standard measures for assessing the accuracy of remotely sensed data known as the overall accuracy and the kappa index [24].

The assessment results of the LULC classification for 2000, 2005, 2010, 2015, and 2020 are shown in Tables below. Sixty reference points were used to evaluate the accuracy of the created land use land cover maps. The overall classification accuracy obtained from the error matrix is shown below each table, and the result of Kappa statistic (**Tables 2–6**).

In conclusion, supervised classification was used because the operator can detect errors and correct them although it is time-consuming.

2.5 Data analysis and presentation methods

Spatial statistics is the collection of statistical methods in which spatial locations play an explicit role in the analysis of data [20]. Most often, spatial statistics are used to detect, characterize, and make inferences about spatial patterns, primarily in ecology

Class name	Number of reference pixels	Number of classified pixels	Correctly classified pixels	User's accuracy	Producer's accuracy
Cropland	13	9	8	88.89	61.54
Bare land	7	6	4	66.67	57.14
Water	11	12	11	91.67	100
Grassland	9	12	7	58.33	77.78
Woodland	12	13	10	76.92	83.33
Built up land	8	8	6	75	75

Overall classification accuracy 77% Kappa statistic 0.718.

Table 2.
Accuracy assessment results for 2000.

Class name	Number of reference pixels	Number of classified pixels	Correctly classified pixels	User's accuracy	Producer's accuracy
Cropland	11	9	7	77.78	63.64
Bare land	9	11	7	63.64	77.78
Water	10	10	10	100	100
Grassland	13	13	8	61.54	61.54
Woodland	9	10	7	70	77.78
Built up land	8	7	5	71.43	62.5

Overall classification accuracy 73% Kappa statistic 0.679.

Table 3.
Accuracy assessment results for 2005.

Class name	Number of reference pixels	Number of classified pixels	Correctly classified pixels	User's accuracy	Producer's accuracy
Cropland	10	11	7	63.64	70
Bare land	12	10	8	80	66.67
Water	8	8	7	87.5	87.5
Grassland	9	11	6	54.55	66.67
Woodland	14	11	10	90.91	71.43
Built up land	7	9	5	55.56	71.43
Overall classification accuracy 72% Kappa statistic 0.659.					

Table 4.
Accuracy assessment results for 2010.

Class name	Number of reference pixels	Number of classified pixels	Correctly classified pixels	User's accuracy	Producer's accuracy
Cropland	10	9	6	66.67	60
Bare land	8	12	6	50	75
Water	9	10	8	80	88.89
Grassland	13	13	9	69.23	69.23
Woodland	11	9	8	88.89	72.73
Built up land	9	7	5	71.43	55.56
Overall classification accuracy 70% Kappa statistic 0.639.					

Table 5.
Accuracy assessment results for 2015.

Class name	Number of reference pixels	Number of classified pixels	Correctly classified pixels	User's accuracy	Producer's accuracy
Cropland	11	7	7	100	63.64
Bare land	12	18	12	66.67	100
Water	9	9	8	88.89	88.89
Grassland	8	10	8	80	100
Woodland	10	10	10	100	100
Built up land	10	6	6	100	60
Overall classification accuracy 85% Kappa statistic 0.819.					

Table 6.
Accuracy assessment results for 2020.

and geography. Spatial patterns can be identified using logistic regression analysis [25]. This method measures the mean nearest distance for all points and assumes all points in the study area have been surveyed. Then, the observed mean distance is compared to the expected mean distance under the null hypothesis that the distribution of points is random. Regression is the determination of a statistical relationship between two or more variables. In simple regression there is only two variables independent and dependent variable, Independent Variable is the cause of the behavior of another one [26]. Regression analysis in the R package was used. The relationship between two variables may be one of the functional dependences of one on the other. For change detection analysis, the post-classification change detection technique was adopted. Data was collected and fed into the R package, whereby simple data analysis tools such as frequencies and percentages were used to analyses data. The presentation was done using R, Microsoft Word, and Excel packages in the form of tables, figures, and charts. QGIS 3.4. Logistic regression analysis in R studio was used for analysis to find if there is a significant relationship between LULC change and Rhino distribution. The prediction of LULC map by 2030 was prepared in 3 stages that is: (i) preparation of datasets/raster layer (LULC layers for 2000, 2010, and 2020), and spatial layers using distance from rivers, and DEM layers.(ii) Training model algorithm using MOLUSCE plugin in GIS version 2.8.(iii) Running simulated model to obtain the LULC state for 2030.

A binomial logistic regression model was used to determine and analyze the impact of LULCC on black rhino distribution in MBRC. Logistic regression analysis was performed in R version 4.0.2. LULC was used as an explanatory variable in the logistic regression model. The results of the model indicated that LULC changes have a significant positive relationship with significant impact on black rhino distribution ($p = 0.0381$). The logistic regression expression used takes the form;

$$P = \frac{1}{1 + e^{-1(\beta 0 + \beta 1X1)}} \quad (1)$$

Where: β_0 = Constant = 0.4677.
β_1 = LULC intercept = 0.04570.
X_1 = LULC map.

3. Results

3.1 Land-use and land cover change

The objective of classification was to group together a set of observational units on the basis of their common attributes. The baseline data for land cover and land use showing 6 different classes, bare land, water body, built-up land, cropland, woodland, and grassland as described in the legend in **Figure 3** were classified (**Tables 7** and **8**) (**Figure 4**).

To compare the LULC percentage change within the period of 20 years from 2000 to 2020, **Figure 5** above was generated and it showed that from 2000 to 2020 there were increments in percentage cover of the bare lands, water, and croplands classifications while woodland, grasslands and built up land went down. The classification that showed a big percentage cover change was the bare land which increased by slightly over 160% followed by cropland with about 73%. Woodland decreased by

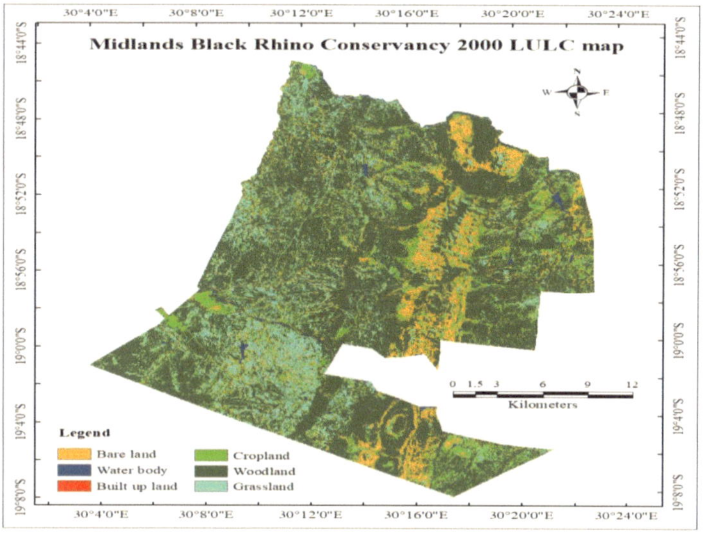

Figure 3.
MBRC LULC map for the year 2000.

Class name	Description
Bare land	This represented virgin land, unoccupied land, and deforested land.
Waterbody	This class included all the dams and other bodies containing clear open water .
Built-up land	These were mainly homesteads.
Cropland	These were mainly planted commercial farms.
Woodland	Included the natural vegetation and shrubs
Grassland	Represented all the areas covered by tall grass and bushes of all types .

Table 7.
Land cover classes.

LULCC(%)	2000	2005	2010	2015	2020
Cropland	13.2	20.75	20.28	10.92	21.56
Water	0.07	14.02	14.29	15.34	20.29
Bare land	12.54	18	23.65	25.98	30.66
Grassland	14.23	10.49	10.91	10.02	7.01
Woodland	35.46	21.1	18.03	20.68	12.17
Built up land	24.5	15.64	12.84	17.06	8.31

Table 8.
Land use land cover change (LULCC) percentage.

about 46% from 2000 to 2020 while built up land also went down by about 39% within the same time. Grasslands percentage cover decreased by about 30% while the water bodies percentage cover increased by over 200%.

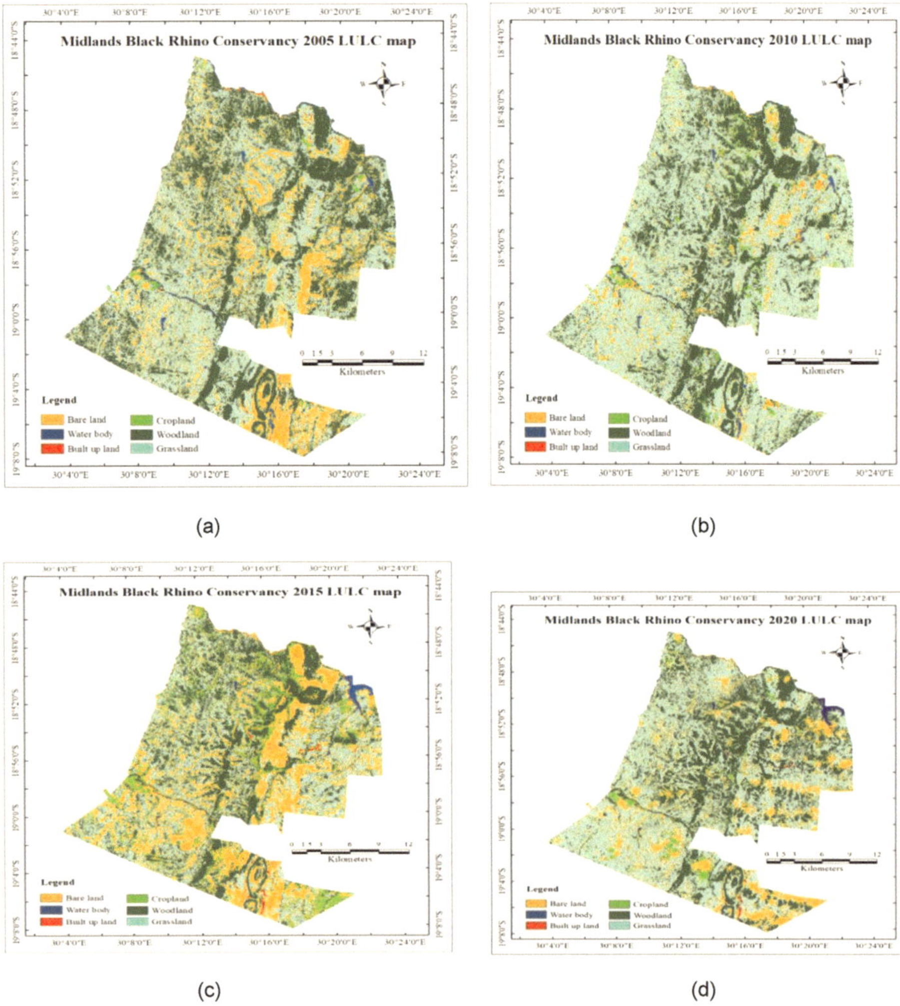

Figure 4.
MBRC LULC map for the year 2005, 2010, 2015 and 2020.

3.2 LULCC, and black rhino distribution

3.2.1 Hypothesis testing

The results of logistic regression analysis were used to test the hypothesis. LULCC may be the major cause of the present pattern of rhino distribution in MBRC, as supported with the statistical analyses ($p = 0.0381$). Rhino distribution (**Figure 6a**) also illustrates the distribution showing dispersed type of distribution which might be, as a result of browse of woodlands evenly distributed in the southern side of the area in the year 2000. However, possibly in the year 2020 due to LULCC the distribution changed showing Black Rhinos clustered in areas where there is dense woody vegetation and river streams (**Figure 6b**). The black rhino distribution in the MBRC area shows a preference for the wooded land. Factors contributing to this preference may

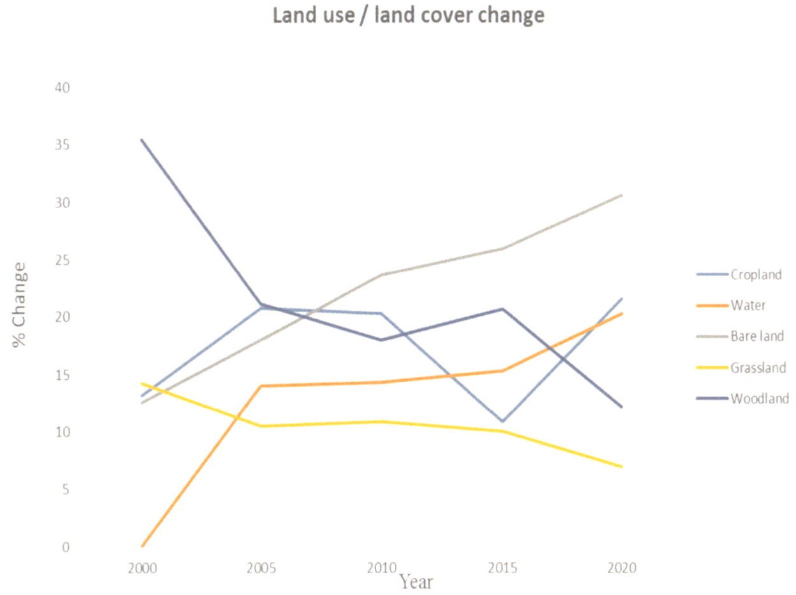

Figure 5.
2000–2020 percentage cover.

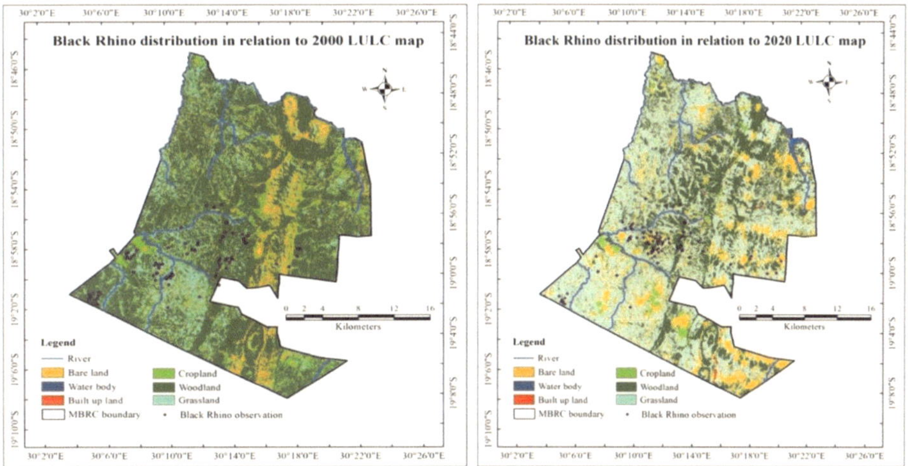

(a) (b)

Figure 6.
Black Rhino distribution for the year 2000 and 2020.

be availability of food, cover, and water. Having said that, interspecific competition can play a role in animal distribution [27], not only through competition for resources, as seen in Black Rhino [28] and African elephant [29], but also by the presence of a physically dominating species [30].

Therefore, the hypotheses that, there is no significant impact of LULC change on Black Rhino distribution in MBRC was not supported by the results, hence rejecting the null hypothesis (**Figure** 7).

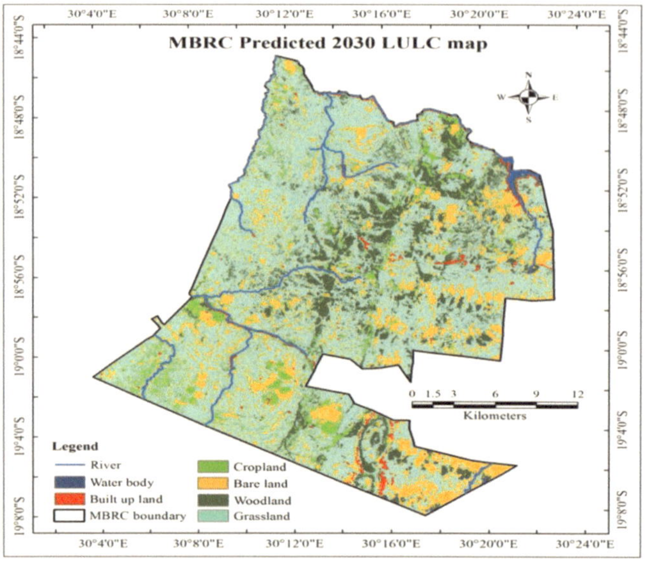

Figure 7.
Predicted map for MBRC by 2030.

Class name	Area (square km)	Percentage
Cropland	60.3585	8.43
Bare land	118.7225	16.59
Water	4.5378	0.63
Grassland	434.2626	60.67
Woodland	87.5151	12.23
Built up land	10.4176	1.46
Total Area 715.8141 square kilometers.		

Table 9.
Predicted 2030 LULC details.

3.3 Prediction of LULCC map by 2030

Prediction of LULCC for the year 2030 was performed in QGIS 2.8 using MOLUSCE plugin. Based on the statistics for the year 2000, 2010, and 2020 it was possible to predict LULCC by 2030.The results predicted an increase in cropland by 8.43%, bareland 16.59%, water 0.63%, grassland 60.67%, woodland, 12.23%, and built up 1.46% as shown in **Table 9** above.

4. Discussion

Damaging human activity continues to encroach on natural environments, thereby destroying the habitats of countless species. As our numbers rise, infrastructure and cropland are growing and merging into each other, fragmenting the remaining habitat and leaving isolated patches for natural populations of plants and animals too small to survive. The increase in bare land is as a result of mining that was done in the area. Mining contributed to lots of deforestation. Furthermore, water cover increased as a result of the open cast mines that were left unclosed hence holding water. Significant increase in cropland experienced may have been triggered by the land reform program since new settlers started to prepare more crop land during the years 2000 to 2005. Agriculture used to be subsistence farming only but, it is no longer subsistence since the farmers have extended the cropland for a better yield. As a result of these anthropogenic factors, land use change has significantly altered the habitat for rhinos in MBRC.

5. Conclusion and recommendations

Quantum GIS and R studio software's made it possible for the formulation of the maps showing the spatial and temporal distribution of Black Rhino furthermore, LULC changes for the month of May year:, 2000, 2005, 2010, 2015, 2020, and modeling of LULCC by 2030 were also made possible. The results predicted an increase in cropland by 8.43% bare land 16.59%, water 0.63%, grassland 60.67%, woodland, 12.23%, and built up 1.46%. Taken together the results, indicated that, LULC changes were significantly impacting Rhino distribution for the period 2000, and 2020. In the year 2000 Rhinos where randomly distributed to the western side of the conservancy however, in the year 2020 Rhinos where found to be clustered in the middle of the conservancy. Furthermore, there is a significant expansion of bare land, and cultivated land noticed. On the other hand there is decrease in woodland area, grasslands area, and built up land area. This study clearly indicated the significant impact of LULC change on Black Rhino distribution. The study also proves that integration of GIS and remote sensing technologies is effective tool for LULCC mapping and modeling. The quantification of LULC changes of MBRC is very useful for environmental management groups, policy makers, and for the academia to better understand the ecosystems.

The researcher recommends constant monitoring of LULCC, and further researches of Black Rhino habitat suitability in MBRC from the management board. The Ministry of Environment, Climate, Tourism, and Hospitality Industry must consider enforcing policies that restricts human encroachment into protected areas set aside for wildlife conservation. The Conservator must introduce re-afforestation of hills and management of woodlands, and grasslands involving both conservation of existing patches of the woodlands and enrichment of degraded grasslands including stream buffering. By linking the cause-effect relations revealed by the analyses, land management prescriptions should be developed for major land use categories in the area. Suitable land use management practices for agricultural land uses should be identified mainly based on their ability to mimic the forest ecosystems.

Author details

Blessing M. Mugaviri[1*], Gilbert H. Moyo[1], Ezra Pedzisai[2] and Cuthbert Maravanyika[1]

1 Scientific Services, Parks and Wildlife Management Authority, Sebakwe Research Station, Kwekwe, Zimbabwe

2 Department of Geography, Bindura University of Science Education, Bindura, Zimbabwe

*Address all correspondence to: mugaviriblessing@icloud.com

References

[1] Lo SO, Tucker CJ, Anyamba AC, Collatz MGJ, Giglio L, Hall FG, et al. Environmental Modelling with GIS and RS. London: Taylor & Francis; 2002

[2] Blaschke T. Object based image analysis for remote sensing. ISPRS Journal of Photogrammetry and Remote Sensing. 2010;**65**:2-16

[3] Blaschke T. Object based image analysis for remote sensing ISPRS Journal of Photogrammetry and Remote Sensing. (Chipunza, TMAP). 2016;**65**:2-16

[4] Armenteras D, Murcia U, González TM, Barón OJ, Arias JE. Scenarios of land use and land cover change for NW Amazonia: Impact on forest intactness. Glob. Ecol. Conserv. 2019;**17**:e00567

[5] Negassa MD, Mallie DT, Gemeda DO. Forest cover change detection using geographic information systems and remote sensing techniques: A spatio-temporal study on Komto protected forest priority area, east Wollega zone, Ethiopia. Environmental Systems Research. 2020, 2020;**9**(1):1-4

[6] Mishra PK, Rai A, Rai SC. Land use and land cover change detection using geospatial techniques in the Sikkim Himalaya, India. EJRS. 2019;**23**:133-143. DOI: 10.1016/j.ejrs.2019.02.001

[7] Belay T, Mengistu DA. Land use and land cover dynamics and drivers in the Muga watershed, upper Blue Nile basin, Ethiopia. RSASE. 2019;**15**:100249

[8] Joshi C, De Leeuw J, van Duren IC. Remote sensing and GIS applications for mapping and spatial modeling of invasive species. In: Proceedings of ISPRS. Vol. 35. Jul 2004. p. B7

[9] Hayes EB. Zika virus outside Africa. Emerging Infectious Diseases. 2009;**15** (9):1347

[10] Saha AK, Gupta RP, Sarkar I, Arora MK, Csaplovics E. An approach for GIS-based statistical landslide susceptibility zonation—with a case study in the Himalayas. Landslides. 2005;**2**(1):61-69

[11] Buk KG, Knight MH. Habitat suitability model for black rhinoceros in Augrabies falls National Park, South Africa. South African Journal of Wildlife Research-24-month delayed open access. 2012;**42**(2):82-93

[12] Ganqa NM, Scogings PF, Raats JG. Diet selection and forage quality factors affecting woody plant selection by black rhinoceros in the great fish river reserve, South Africa. South African Journal of Wildlife Research-24-month delayed open access. 2005;**35**(1):77-83

[13] Collis J, Hussey R, Hussey J. Business Research: A Practical Guide to Undergraduate and Postgraduate Students. 2nd ed. Basingstoke: Palgrave Macmillan; 2003

[14] Bowman A, Wyman JF, Peters J. The operations manual: A mechanism for improving the research process. Nursing Research. 2012;**51**(2):134-138

[15] Zimbabwe Ministry of Water Resources and Development. 2004. Available from: http://worldcat.org/identities/lccn-n84129684/

[16] McAlpine CA, Fensham RJ, Temple-Smith DE. Biodiversity conservation and vegetation clearing in Queensland: Principles and thresholds. The Rangeland Journal. 2002;**24**:36-55. DOI: 10.1071/RJ02002

[17] Burns N, Grove SK. The Practice of Nursing Research. 5th ed. Philadelphia: W.B. Saunders Company; 2014

[18] Help RUS. Developing an effective dissemination plan. 2011. Available from: www.researchutilization.org/matrix/resources/index.html [Accessed: December 6, 2011]

[19] Munro BH. Statistical Methods for Health Care Research. 5th ed. Philadelphia: Lippincott; 2014 Q1

[20] Moher D. CONSORT: An evolving tool to help improve the quality of reports of randomized controlled trials. Consolidated Standards of Reporting Trials. JAMA. 2018;**279**(18):1489-1491

[21] Wasserberg G, Kotler BP, Abramsky Z. The role of site, habitat, seasonality and competition in determining the nightly activity patterns of psammophilic gerbils in a centrifugally organized community. Oikos. 2006;**112**:573-579

[22] Ribeiro P Jr, Peter D. GeoR: A package for Geostatistical analysis. R-NEWS. 2001;**1**:14-18

[23] Fernandez-Gallego J, Ma B, Aparicio Gutiérrez N, Nieto-Taladriz M, Araus J, Kefauver S. Automatic wheat ear counting using thermal imagery. Remote Sensing MDPI AG. 2019;**11**:751

[24] Jensen JR. Introductory Digital Image Processing: A Remote Sensing Perspective. 3rd ed. Upper Saddle River: Prentice-Hall; 2005. p. 526

[25] Congalton R, Green K. Assessing the Accuracy of Remotely Sensed Data: Principles and Practices. Boca Raton: CRC/Lewis Press; 1999. p. 137

[26] Fortin M-J, Payette S. How to test the significance of the relation between spatially autocorrelated data at the landscape scale: A case study using fire and forest maps. Écoscience. 2002;**9**:213-218

[27] Lincoln YS, Guba EG. Naturalistic Inquiry. Newbury Park, CA: Sage Publications; 2015

[28] Garshelis DL. Delusions in habitat evaluation: Measuring use, selection, and importance. In: Boitani L, Fuller TK, editors. Research Techniques in Animal Ecology. New York: Controversies and Consequences. Columbia University Press; 2000. pp. 111-164

[29] Birkett A. The impact of giraffe, rhino and elephant on the habitat of a black rhino sanctuary in Kenya. African Journal of Ecology. 2002;**40**:276-282

[30] Kerley GIH, Landman M. The impacts of elephants on biodiversity in the eastern cape subtropical thickets. South African Journal of Science. 2006; **102**:395-402

Chapter 5

Floristic and Ethnobotanical Study of Indigenous Plants of Ranapur Reserve Forest, Odisha, India

Diptiman Sahoo, Gyanesh Dash, K.T.K.G. Ranjan Mohanty, Srinivas Acharya, Ehsan Amiri Ardakani, Monali Priyadarsini Mishra and Gyanranjan Mahalik

Abstract

The indigenous plants of Ranapur reserve forest of Odisha state, India, possess ample scope in ethnobotany due to their medicinal properties. Keeping in view of the requirements of the urbanites, this work highlights the utilization of native flora, identification, ethnobotany, floristic survey, and preservation of natural and cultivated plant species within the hills of Ranapur, Odisha. A field study around the Ranapur forest resulted in a wide diversity accounting for 143 plant species belonging to 53 families, which were identified and documented alongside their botanical name, family, and habitat following local herbaria, archives of Flora of Orissa, monographs, and standard taxonomic study. *Gmelina arborea*, *Rauwolfia serpentina*, and *Crataeva nurvala* (Varuna) fall into the RET taxa (Rare, Endangered, and Threatened) and are groundbreaking against various diseases. A few wild plants such as *Shorea robusta*, *Dalbergia sissoo*, *Pterocarpus marsupium, Murraya koenigii*, and *Schleichera oleosa* were the most dominant species in the study area. The present study adds detailed data-base concerning the floral diversity and their medicinal values, which attracts many researchers as well as the local populace to conserve and explore their wide-spectrum applications. This could be useful in novel drug discovery and authenticates the ethnomedicinal knowledge.

Keywords: ethnobotanical, floristic survey, medicinal, RET, Ranapur, traditional

1. Introduction

Study on bionomics and the allocation of organisms within an abundant elevation slope have been considered in numerous current projects in mountainous ecosystems across the globe [1, 2], particularly with an emphasis on flora and species attributes [3]. From the beginning of human race, the humans have been dependent on plants for their different needs. Plants provide food, shelter, and medicine. It is estimated that around 10 million species of plants occupy the earth, of which, nevertheless 1.7 million species are acknowledged to science [4]. The Ranapur reserve forest has been the home of various tribal communities. The most dominated tribes are Kondha and Soura, whereas other sub-tribes are also found in the interior forest sub-tribe such as Relli, Jhodia, Dora, Kotia, etc. These tribes depend on traditional medicinal plants for their health care [5].

Besides, information on plant species and their decent variety inside the long slope is essential for any environmental and vegetation investigation, especially for inadequately known zones. With these devices, it is conceivable to grow protection and the management exercises in every territory [6]. The current study is a floristic and ethnobotanical survey of Maninag hill located in Ranapur reserve forest area of Nayagarh district of Odisha. The objectives of the study were collection and identifications of plants in the Maninag hill to study medicinal status of the area, documentation of traditional knowledge, and use of medicinal plants for well-being, to enumerate the collected plant species, to study the distribution pattern of vegetation in the study area.

2. Materials and methods

2.1 Study site

The study area is situated between 20.0628°N, 85.3433°E in Nayagarh district, Odisha (**Figure 1**). Raj Ranapur is a modest community in the region of Nayagarh in the eastern Indian state of Odisha. The town is otherwise called Ranapurgarh or essentially Ranapur according to the cutting-edge use. The town is truly huge particularly during the British Raj when it was the capital of the regal state of Ranapur. The Ranapur town is situated in the lower regions of Maninag Hills, which is a hill framework covering the entire of Ranapur and a significant part of the encompassing territories. It is one of the most significant towns in the Nayagarh district and furthermore one of the significant places in Odisha [7].

Figure 1.
Map of study area (source: compare info base private Ltd., 2003).

2.2 Methodology

Extensive and intensive field surveys were conducted in different seasons to explore the floristic composition and to collect ethnobotanical information. The plant specimens are collected and photographs were also taken from the study area. During this period, interviews with temple priest and local inhabitants were conducted to collect information about the medicinal use of different plant specimens with their vernacular names. Maninag hills being a natural sacred site, religious beliefs, spirituality, and the participation of locals on conservation of this site were also documented [8, 9].

2.3 Plant collection and herbarium studies

The supportive plant specimens were collected, processed, critically studied, identified, and preserved in the Herbarium. Different Herbaria of Bhubaneswar that held the specimens of earlier workers were visited and checked their identity. Voucher specimens were identified by referring standard local floras [10–12].

3. Results

The plants seen in the study region have been listed and documented in the accompanying table with their botanical names, family, and local names. The plants were observed and their development propensities have additionally been recorded. During the investigation, 143 plant species belonging to 53 families were recorded from the study area, i.e., Maninag Hill of Ranapur area, Nayaghar, Odisha (**Tables 1** and **2**). Habit-wise analysis of the available species indicated that 33 (23%) were herbs followed by 57 (40%) trees, 22 (15%) climbers, 21 (15%) shrubs, 10 (7%) grasses (**Figure 2**). Family-wise trend in diversity of species dominance followed a pattern of family Fabaceae (9), Poaceae (7), Apocynaceae (7), Mimosaceae (7), Asteraceae (7), Caesalpiniaceae (7), and Convolvulaceae (6) (**Figure 3**). Out of 143 plants, a few important medicinal plant species were observed during the study, i.e., *Hemidesmus indicus*, *Andrographis paniculata* and *Tinospora cordifolia*, *Vetiveria zizanioides*, *Terminalia bellirica*, *Terminallia chebula*, *Nyctanthes arbor-tristis*, *Lawsonia inermis*, *Justicia adhatoda*, *Phyllanthus fraternus*, *Tridax procumbens*, *Desmotachya bipinnata*, etc., which were used by tribes such as Gond, Khaira, Kolho, Koya, Lodha, Malua, Kandha, Santal, Sabar, etc., to cure diseases such as wound healing, skin diseases, diarrhea, jaundice, urinary tract infections, stomach problems, etc. There are three plants found in this study area that are categorized as RET (Rare, Endangered, and Threatened), i.e., *Gmelina arborea*, *Rauwolfia serpentina*, and *Crataeva nurvala*.

4. Discussion

The plant diversity of a locale is the aggregate of the species within its boundaries, regardless of whether wild or cultivated, which is a reflection of vegetation and plant resources. Plant resources are influenced by agriculture, overgrazing, anthropogenic activities, and catastrophic events. Investigation and checking of biodiversity of any region are essential for the preservation and the board arranging. This investigation

Sl. no.	Scientific name	Family	Local name
1.	*Acacia auriculiformis* A.Cunn. ex Benth.	Mimosaceae	Akashi
2.	*Acacia leucophloea* (Roxb.) Willd.	Mimosaceae	Gohira
3.	*Acacia catechu* (L.f.) Wild.	Mimosaceae	Khaira
4.	*Acacia nilotica* (L.) Wild ex. Del.	Mimosaceae	Babul
5.	*Aegle marmelos* (L.) Corr.	Rutaceae	Bela
6.	*Albizia lebbeck* (L.) Benth	Mimosaceae	Sirisa
7.	*Alstonia scholaris* (L.) R.Br.	Apocynaceae	Chhatiana
8.	*Anacardium occidentale* L.	Anacardiaceae	Kaju
9.	*Annona squamosa* L.	Annonaceae	Aata
10.	*Artocarpus heterophyllus* Lam.	Moraceae	Panasa
11.	*Azadirachta indica* A.Juss	Meliaceae	Nimba
12.	*Bauhinia variegate* L.	Caesalpiniaceae	Kanchana
13.	*Bombax ceiba* L.	Bombacaceae	Simili
14.	*Borassus flabellifer* L.	Arecaceae	Tala
15.	*Buchanania lanzan* Spreng.	Anacardiaceae	Chara
16.	*Butea monosperma* (Lam.) Taub.	Fabaceae	Palasa
17.	*Cassia siamea* (Lam.) H.S. Irwin & Barneby	Fabaceae	Chakunda
18.	*Cassia fistula* L.	Caesalpiniaceae	Sunari
19.	*Casuarina equisetifolia* L.	Casuarinaceae	Jhaun
20.	*Ceiba pentandra* (L.) Gaertn.	Bombacaceae	Sweta Simili
21.	*Chloroxylon swietiana* DC.	Rutaceae	Bheru
22.	*Cocos nucifera* L.	Arecaceae	Nadia
23.	*Dalbergia sissoo* Roxb.	Fabaceae	Sisu
24.	*Delonix regia* (Boj. Ex Hook) Raf.	Caesalpiniaceae	Kruchnachuda
25.	*Dillenia indica* L.	Dilleniaceae	Oou
26.	*Diospyros malabarica* (Desr.) Kostel.	Ebenaceae	Mankadalendu
27.	*Diospyros melanoxylon* Roxb.	Ebenaceae	Kendu
28.	*Eucalyptus citriodora* Hook.	Myrtaceae	Nilagiri
29.	*Ficus benghalensis* L.	Moraceae	Bara
30.	*Ficus racemosa* L.	Moraceae	Dimiri
31.	*Ficus religiosa* L.	Moraceae	Asta
32.	*Haldina cordifolia* (Roxb.) Ridsd.	Rubiaceae	Kurum
33.	*Holarrhena pubescens* (Buch-Ham.) Wall. Ex. G.Don.	Apocynaceae	Kuruchi
34.	*Lagerstroemia reginae* Roxb.	Lythraceae	Patuli
35.	*Leucaena leucocephala* (Lam.) de Wit.	Mimosaceae	Nagarjuna
36.	*Madhuca indica* Gmel.	Sapotaceae	Mahula
37.	*Mallotus phillippensis* (Lam.) Muell.	Euphorbiaceae	Sinduri
38.	*Mangifera indica* L.	Anacardiaceae	Amba

Sl. no.	Scientific name	Family	Local name
39.	*Michelia champaca* L.	Magnoliaceae	Champa
40.	*Moringa oleifera* Lam.	Moringaceae	Sajana
41.	*Neolamarckia cadamba* (Roxb.) Benth	Rubiaceae	Kadamba
42.	*Phoenix sylvestris* (L.) Roxb.	Arecaceae	Khajuri
43.	*Pongamia pinnata* (L.) Pierre.	Fabaceae	Karanja
44.	*Psidium guajava* L.	Myrtaceae	Pijuli
45.	*Semecarpus anacardium* L.f.	Anacardiaceae	Bhalia
46.	*Shorea robusta* Gaertn.f.	Dipterocarpaceae	Sala
47.	*Simarouba glauca* DC.	Simaroubaceae	Simarouba
48.	*Strychnos nux-vomica* L.	Loganiaceae	Kochila
49.	*Streblus asper* Lour.	Moraceae	Sahada
50.	*Syzygium cumini* (L.) Skeels.	Myrtaceae	Jammu
51.	*Tamarindus indica* L.	Caesalpiniaceae	Tentuli
52.	*Tectona grandis* L.f.	Verbenaceae	Saguan
53.	*Terminalia alata* Heyne ex Roth.	Combretaceae	Asana
54.	*Terminalia arjuna* (Roxb. ex Dc.) Wight.	Combretaceae	Arjuna
55.	*Terminalia bellirica* (Gaertn) Roxb.	Combretaceae	Bahada
56.	*Terminalia chebula* Retz.	Combretaceae	Harida
57.	*Ziziphus mauritiana* Lam.	Rhamnaceae	Barakoli
Shrubs			
1.	*Calotropis gigantea* R.Br.	Asclepiadaceae	Arakha
2.	*Cascabela thevetia* (L.) Lippold.	Apocynaceae	Kaniar
3.	*Cassia occidentalis* L.	Caesalpiniaceae	Chakunda
4.	*Cassia hirsute* L.	Caesalpiniaceae	Chhota chakunda
5.	*Chromolaena odorata* (L.) King. & Robins.	Asteraceae	Guhia
6.	*Clerodendrum viscosum* Vent.	Verbenaceae	Kharkhari
7.	*Croton roxburghii* Balak.	Euphorbiaceae	Masudi
8.	*Glycosmis pentaphylla* (Retz.) Dc.	Rutaceae	Chauladhua
9.	*Ipomoea carnea* Jacq. spp *fistulosa* (Mart. ex Choisy) Austin	Convolvulaceae	Amari
10.	*Jatropha gossypifolia* L.	Euphorbiaceae	Gaba
11.	*Justicia adhatoda* L.	Acanthaceae	Basanga
12.	*Lantana camara* L.	Verbenaceae	Naguari
13.	*Lawsonia inermis* L.	Lythraceae	Manjuati
14.	*Nyctanthes arbor-tristis* L.	Oleaceae	Gangasiuli
15.	*Pavetta crassicaulis* Bremek.	Rubiaceae	Macharanka
16.	*Phoenix acaulis* Buch.-Ham. Ex Roxb.	Arecaceae	Banakhajur
17.	*Punica granatum* L.	Punicaceae	Dalimba

Sl. no.	Scientific name	Family	Local name
18.	*Rauwolfia tetraphylla* L.	Apocynaceae	Patalagaruda
19.	*Ricinus communis* L.	Euphorbiaceae	Jada
20.	*Vitex negundo* L.	Verbenaceae	Begunia
21.	*Ziziphus oenoplia* (L.) Mill.	Rhamnaceae	Kanteikoli
Climbers			
1.	*Allamanda cathartica* L.	Apocynaceae	Harakara
2.	*Aganosma caryophyllata* (Roxb. ex Sims.)	Apocynaceae	Malati
3.	*Argyreia nervosa* (Burm.f.) Boj.	Convolvulaceae	Bataraj
4.	*Aristolochia indica* L.	Aristolochace	Panairi
5.	*Asparagus racemosus* Willd.	Liliaceae	Satabari
6.	*Bauhinia vahlii* Wight. & Arn.	Caesalpiniaceae	Sialilata
7.	*Butea superb* Roxb.	Fabaceae	Latapalasa
8.	*Clerodendrum speciosum* L.	Verbenaceae	-
9.	*Cucumis sativus* L.	Cucurbitaceae	Pitakakudi
10.	*Dioscorea alata* L.	Dioscoreaceae	Bana-allu
11.	*Dioscorea pentaphylla* L.	Dioscoreaceae	Karaba
12.	*Dioscorea wallichii* Hook.f.	Dioscoreaceae	Pitaallu
13.	*Hemidesmus indicus* (L.) R.Br.	Asclepiadiaceae	Anantamula
14.	*Ichnocarpus frutescens* (L.) R.Br.	Apocynaceae	Saonllar
15.	*Ipomoea cairica* (L.) Sweet.	Convolvulaceae	Banakalama
16.	*Ipomoea quamoclit* L.	Convolvulaceae	Kujalata
17.	*Ipomea pes-tigridis* L.	Convolvulaceae	Billenandi
18.	*Mikania micrantha* Kunth.	Asteraceae	Salamari
19.	*Mucuna pruriens* (L.) DC.	Fabaceae	Baidanka
20.	*Passiflora foetida* L.	Passifloraceae	Pasaruni
21.	*Quisqualis indica* L.	Combretaceae	Madhumalati
22.	*Tinospora cordifolia* (Willd.) Hook.f. & Thomas	Menispermaceae	Guduchilata
Herbs			
1.	*Abutilon indicum* (L.) Sweet	Malvaceae	Pedipedika
2.	*Abelmoschus manihot* (L.) Medic.	Malvaceae	Banabhendi
3.	*Aerva lanata* (L.) Juss.	Amaranthaceae	Paunsia
4.	*Agae americana* L.	Agavaceae	Agara
5.	*Ageratum conyzoides* L.	Asteraceae	Pokasunga
6.	*Andrographis paniculata* (Burm.f.) Wall.ex Nees	Acanthaceae	Bhuinlimba
7.	*Atylosia scarabaeoides* (L.) Benth.	Fabaceae	Banakolatha
8.	*Barleria cristata* L.	Acanthaceae	Daskarenta
9.	*Boerhavia diffusa* L.	Nyctaginaceae	Puruni

Sl. no.	Scientific name	Family	Local name
10.	*Blumea lacera* (Burm.f.) DC.	Asteraceae	Kukursunga
11.	*Cajanus cajan* (L.) Huth.	Fabaceae	Banaharada
12.	*Costus speciosus* (Koening) Sm.	Zingiberaceae	Gaigobara
13.	*Crotalaria pallida* Ait.	Fabaceae	Jhun-Junka
14.	*Curculigo orchioides* Gaertn.	Hypoxidaceae	Talamuli
15.	*Curcuma longa* L.	Zingiberaceae	Haladi
16.	*Elephantopus scaber* L.	Asteraceae	Mayurachulia
17.	*Emilia sonchifolia* (L.) DC.	Asteraceae	Sarkara
18.	*Euphorbia hirta* L.	Euphorbiaceae	Chitakutei
19.	*Evolvulus nummularius* (L.) L.	Convolvulaceae	Bichhamali
20.	*Hedyotis corymbosa* (L.) Lam.	Rubiaceae	Gharapodia
21.	*Hedyotis diffusa* Willd.	Rubiaceae	-
22.	*Hybanthus enneaspermus* (L.) F.V.Muell.	Violaceae	Madanmastak
23.	*Hyptis suaveolens* (L.) Poit.	Lamiaceae	Gangatulasi
24.	*Kalanchoe pinnata* (Lam.) Pers.	Crassulaceae	Hemakedara
25.	*Leonotis nepetifolia* (L.) R.Br.	Lamiaceae	Kantasia
26.	*Leucas aspera* (Willd.) Link.	Lamiaceae	Gayasa
27.	*Mimosa pudica* L.	Mimosaceae	Lajakuli
28.	*Ocimum gratissimum* L.	Lamiaceae	Banatulasi
29.	*Phyllanthus fraternus* Webster	Euphorbiaceae	Bhuinonla
30.	*Sida acuta* Burm.f.	Malvaceae	Bajramuli
31.	*Sida cordata* (Burm.f.) Borssum	Malvaceae	Bajramuli
32.	*Solanum nigrum* L.	Solanaceae	Lunlunia
33.	*Tridax procumbens* L.	Asteraceae	Bisalyakarani
Grass			
1.	*Bambusa bambos* (L.) Voss.	Poaceae	Baunsa
2.	*Chloris barbata* Sw.	Poaceae	Gandhi
3.	*Cynodon dactylon* (L.) Pers.	Poaceae	Duba
4.	*Cyperus flabelliformis* Rottb.	Cyperaceae	
5.	*Cyperus iria* L.	Cyperaceae	Swanti
6.	*Cyperus rotundus* L.	Cyperaceae	Mutha
7.	*Desmostachya bipinnata* (L.) Stapf.	Poaceae	Kusha
8.	*Eragrostis cilliaris* (L.) R.Br.	Poaceae	-
9.	*Oryza sativa* L.	Poaceae	Dhana
10.	*Vetiveria zizanioides* (L.) Nash.	Poaceae	Bena

Table 1.
Plant list of Ranapur forest, Odisha.

Sl. no.	Name of the family	No. of species
1.	*Acanthaceae*	3
2.	*Agavaceae*	1
3.	*Amaranthaceae*	1
4.	*Anacardiaceae*	4
5.	*Annonaceae*	1
6.	*Apocynaceae*	7
7.	*Arecaceae*	4
8.	*Aristolochace*	1
9.	*Asclepiadaceae*	1
10.	*Asclepiadiaceae*	1
11.	*Asteraceae*	7
12.	*Bombacaceae*	2
13.	*Caesalpiniaceae*	7
14.	*Casuarinaceae*	1
15.	*Combretaceae*	5
16.	*Convolvulaceae*	6
17.	*Crassulaceae*	1
18.	*Cucurbitaceae*	1
19.	*Cyperaceae*	3
20.	*Dilleniaceae*	1
21.	*Dioscoreaceae*	1
22.	*Dipterocarpaceae*	1
23.	*Ebenaceae*	2
24.	*Euphorbiaceae*	6
25.	*Fabaceae*	9
26.	*Hypoxidaceae*	1
27.	*Lamiaceae*	4
28.	*Liliaceae*	1
29.	*Loganiaceae*	1
30.	*Lythraceae*	2
31.	*Magnoliaceae*	1
32.	*Malvaceae*	4
33.	*Meliaceae*	1
34.	*Menispermaceae*	1
35.	*Mimosaceae*	7
36.	*Moraceae*	5
37.	*Moringaceae*	1
38.	*Myrtaceae*	3
39.	*Nyctaginaceae*	1
40.	*Oleaceae*	1
41.	*Passifloraceae*	1
42.	*Poaceae*	7

Sl. no.	Name of the family	No. of species
43.	*Punicaceae*	1
44.	*Rhamnaceae*	2
45.	*Rubiaceae*	5
46.	*Rutaceae*	3
47.	*Rutaceae*	3
48.	*Sapotaceae*	1
49.	*Simaroubaceae*	1
50.	*Solanaceae*	1
51.	*Verbenaceae*	5
52.	*Violaceae*	1
53.	*Zingiberaceae*	1
	Total	**143**

Table 2.
Taxonomic classification (family-wise distribution) of Ranapur forest division, Odisha.

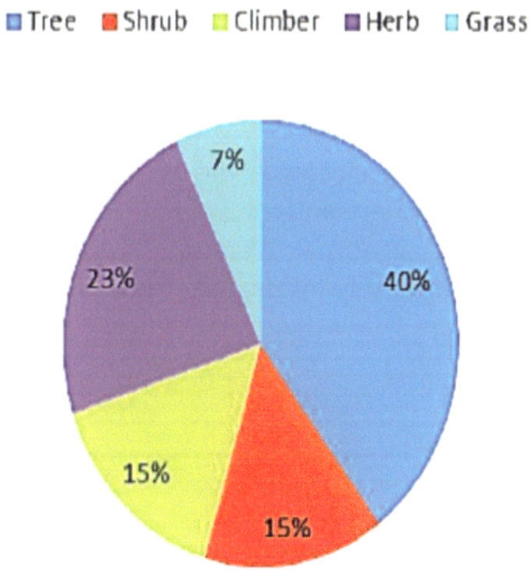

Figure 2.
Floristic diversity of Ranapur forest division, Odisha.

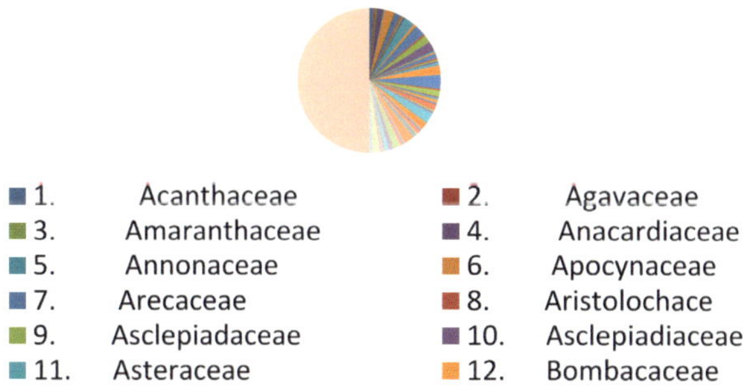

Figure 3.
Family-wise distribution of plant.

reveals that the study area fills in as a wellspring of livelihood for occupying the territory. Ethnobotany is maybe the most significant technique to study natural resources and their administration, which was done by indigenous people since days of yore. It enables us to work with local people to explore knowledge based on experiences of age [13]. Analysis of data reveals various species that have ethnobotanical importance and used for various purposes by the indigenous people of the area. The local herbs particularly medicinal species even today assume a significant job in the financial inspire of the rustic zones and different privately created drugs are as yet being utilized as family unit solutions for different ailments [14–16].

5. Conclusion

Biodiversity is fundamental for human endurance and financial prosperity and for the ecosystem function and stability. The present investigation showed that the Ranapur forest has high species diversity with more than nine different tribes who depend on plants and folk medications. The Fabaceae group of plants are found to be most diverse in the study area. The need of great importance is to aware folks concerning its significance, involvement of people in its.

Acknowledgements

The authors would like to thank the local and traditional healers for sharing their valuable knowledge. All authors are also grateful to HOD, Department of Botany, School of Applied Sciences, Centurion University of Technology and Management (CUTM), Odisha, India, for their help and suggestions.

Conflict of interest

The authors declare that there is no conflict of Interest.

Author details

Diptiman Sahoo[1], Gyanesh Dash[1], K.T.K.G. Ranjan Mohanty[1], Srinivas Acharya[2], Ehsan Amiri Ardakani[3], Monali Priyadarsini Mishra[4] and Gyanranjan Mahalik[1]*

1 Department of Botany, School of Applied Sciences, Centurion University of Technology and Management, Odisha, India

2 Environmental Science Facility, Department of Botany, Utkal University, Bhubaneswar, India

3 Shiraz University of Medical Sciences, Iran

4 School of Paramedics and Allied Health Sciences, Centurion University of Technology and Management, Odisha, India

*Address all correspondence to: gyanranjan.mahalik@cutm.ac.in

References

[1] Beck J, McCain CM, Axmacher JC, Ashton LA, Bärtschi F, Brehm G, et al. Elevational species richness gradients in a hyperdiverse insect taxon: A global meta-study on geometrid moths. Global Ecology and Biogeography. 2017;**26**:412-424

[2] Megersa M, Woldetsadik S. Ethnobotanical study of medicinal plants used by local communities of Damot Woyde district, Wolaita zone, southern Ethiopia. Nusantara Bioscience. 2022;**14**:1

[3] Sanders NJ, Rahbek C. The patterns and causes of elevational diversity gradients. Ecography. 2012;**35**:1-3

[4] Aziz WN, Deshmukh VR, Rothe SP. A floristic survey of flowering plants from Vidyabharati Mahavidyalya Campus, Amravati (Maharashtra) India. International Journal of Life Sciences. 2015;**3**:249-254

[5] Mishra J, Behera SK, Panigrahi AK, Sahu RK, Leelaveni A. Study of Some Ethnomedicinal Plants of Nayagarh District, Odisha. The Journal of Ethnobiology and Traditional Medicine Plants. 2015;**124**:187-199

[6] Jafari SM, Akhani H. Plants of jahan nama protected area, Golestan province. Pakistan Journal of Botany. 2008;**40**:1533-1554

[7] Gazetteer I. Imperial gazetteer of India. Provincial Series: Punjab. 1908;**1**:195

[8] Jain SK. Some Magico-religious Beliefs about Plants among Adibasis of Orissa. Orissa: Tribal Research Bureau; 1971

[9] Jain SK. Dictionary of Indian Folk Medicine and Ethnobotany. New Delhi, India: Deep Publications; 1991

[10] Saxena HO, Brahmam M. The Flora of Orissa (1994-96). Vol. 1-4. Bhubaneswar: Regional Research Laboratory and Forest Development Corporation of Orissa; 1994

[11] Haines HH. The Botany of Bihar and Orissa, 6 parts, London (1921-25). Calcutta: Botanical Survey of India; 1921

[12] Dash G, Mohanty KKGR, Sahoo D, Mahalik G, Parida S. Traditional medicinal plants used for the treatment of asthma in Bhubaneswar, Odisha. International Journal of Herbal Medicine. 2018;**6**:57-60

[13] Behera K, Mandal U, Panda M, Mohapatra M, Mallick SK, Routray S, et al. Ethnobotany and folk medicines used by the local healers of Bhadrak, Odisha. India. Egyptian Journal of Botany. 2021;**61**(2):375-389. DOI: 10.21608/EJBO.2020.26337.1474

[14] Qureshi RA, Ghufran MA. Medicinal Value of Some Important Roses and Allied Species of Northeren Area of Pakistan. Islamabad: Pictorial Printers (Pvt.). Ltd; 2005

[15] Lawal IO, Rafiu BO, Ale JE, Majebi OE, Aremu AO. Ethnobotanical survey of local flora used for medicinal purposes among indigenous people in five areas in Lagos State, Nigeria. Plants. 2022;**11**(5):633

[16] Debidutta R, Mohanta MR, Sahu SC. Floristic diversity of climbing plants in tropical forests of Similipal Biosphere Reserve, Odisha, India. Notulae Scientia Biologicae. 2022;**14**(1):11003

Chapter 6

Characteristics of the Tropical Hardwood – Tree Species for Renewable Energy Production in Zambia

Obote Shakacite, Phillimon Ngandwe, Vincent Nyirenda and Donald Chungu

Abstract

This work studied the diversity, abundance, and distribution of 25 most suitable indigenous tropical hardwood tree species for value-added renewable energy production. The study aimed to assess relative abundance, diversity, distribution, current status, and uses by local communities of these species which are still poorly known. The study is based on data from nationwide remeasured permanent sample plots for Zambia covering different types of forests in agro-ecological zones 1 and 2. Diameter at breast height (DBH) $\geq$2.5 cm was collected and analyzed in all plots. The study approach included informant interviews that focused on species' uses and their availability in the surrounding forests and woodlands and species population inventory of the natural forests and woodlands. The tree species for renewable energy production were determined. The species were identified on the basis of abundance, diversity, regeneration status, and perceived utilization. The study observed the need to manage hardwood supply sustainably, promote lesser-known hardwood tree species, and diversify their use in the wood industry. The intervention of government and other stakeholders to tackle wood fuel production problem using collaborative approach is emphasized.

Keywords: abundance, anthropogenic activities, distribution, diversity, regeneration, species

1. Introduction

Forests and woodlands in Africa play an important role in the livelihood of many rural communities and in the economic development of many developing countries [1]. Globally, it is estimated that 52% of the total forests are in tropical regions, and they are known to be the most important resource in terms of bioenergy production [2]. Tropical hardwoods are important for supplying wood fuel and timber [3, 4]. Tropical hardwoods are preferred owing to a number of characteristics that set it apart from other tree species. Some of the characteristics among species that influence commercial importance include wood color, energy, density, durability, and grain texture. The hardwood species are also influenced by economic factors (availability, wood properties, and market acceptance). *Baikiaea plurijuga* is an important species in Zambia as it is a source of hardwood timber used for railway sleepers, furniture, and flooring [5]. The wood is heavy, with a density of 800–950 kg/m^3 at 12% moisture content. Another important hardwood species in Zambia preferred for its characteristics is *Guibourtia coleosperma*. The wood is heavy, with a density of 670–960 kg/m^3 at 12% moisture content. The attractive appearance of this wood influences its commercial importance as it is widely used in construction, flooring, and joinery among others [5]. The local community had been traditionally using specific hardwood tree species for fuel wood for the generation of energy. Some hardwood species such as *B. boehmii*, *Brachystegia spiciformis* Benth., *J. globiflora*, *Parinari curatellifolia* Planch. ex Benth, and *Uapaca kirkiana* Müll. Arg are mainly used for fuel by local communities [6]. The literature on characterization of properties of hardwood tree species in Southern Africa has expanded [7]; however, there are still limited studies on these species [8]. There is lack of scientific data to validate their preferences of these species in terms of determination of their density, volatile matter, ash, moisture, and fixed carbon content [9].

In Zambia, wood is one of the most abundant natural resource, but there has been limited investment in the wood industry [10]. The country has the largest forest resource in Southern Africa despite its forest cover declining over the years from over 66% (49.9 million ha) in 2008 to 58.7% (44.1 million ha) in 2016, and the current total growing stock has been estimated at 3.2 billion m^3 of which 43% is considered of commercial importance [10]. Some of these forests are managed as protected areas for future development of the wood industry. The demand for wood industry products is increasing paving way for productivity improvement and further development of the wood sector. The current wood consumption is estimated to be 201 million m^3/annum, with almost 98% of this wood originating from natural forests. Zambia consumes about 8.1 million m^3/annum of wood for wood charcoal, and 9.7 million m^3/annum is mainly used for cooking in rural and peri-urban areas. Traditional charcoal production, a major source of employment for the rural poor, relies on the traditional earth kiln which is blamed to be a major contributor to preferred tree species loss in many rural and urban regions of sub-Saharan Africa. The traded wood fuels are used by 70–90% of the 50 million urban population in the region [11, 12] and support a significant flow of money from urban to rural population mainly in areas producing charcoal. The woodlands are a major source of biomass fuel (firewood and charcoal) for household consumption and income for a large proportion of rural livelihoods [13]. Earlier studies have shown that charcoal business is a lucrative local economy, estimated to generate an annual income of about US$ 350 million in Tanzania [14, 15], US$ 30 million in Zambia [13], average annual income of about US$ 250–300 per family in Mozambique [4], and equally in Malawi and Kenya many people owe their livelihoods to the charcoal business [16]. Global charcoal production which generates income for many people is expected to increase despite lack of regulation means and inefficiency in its production [17].

Wood fuel issues in Zambia are increasingly becoming urgent given that wood fuel provides for about 90% of domestic energy needs both in urban and rural areas; hence, demand is rising [18]. Wood fuel production and trade may be blamed for woodland loss where charcoal production from open and protected forest areas alone is shown to have contributed to about 25% of forest loss. Therefore, it is important to understand abundance, diversity, and distribution of the most preferred hardwood tree species for wood fuel production and enhance sustainability. However, any strong dependence on wood fuel for domestic energy consumption in Zambia increasingly puts pressure on production areas. However, the miombo woodland species possesses good potential for recovery through proper care of natural regeneration. There is considerable potential to introduce alternatives to fuel wood and wood charcoal as well as expanding commercial forest plantation for industrial wood. The estimated wood wastes (including harvested and mill wastes) produced on an annual basis from miombo woodland is about 600,000 m^3 which offers a huge opportunity to be developed into fuel wood. According to [10], the surplus is potential supply from natural woodlands that provide a sustainable avenue for future expansion of wood products industry.

Renewable energy is considered an important resource virtually in every aspect of the economic and social development in many countries around the world [19], but globally less than 15% of primary energy supply is renewable energy, and the major part is hydropower and wood fuels in developing countries. The major drivers of wood fuel consumption are population growth, rapid urbanization, poverty, and lack of income growth [17]. In Africa, most countries still depend heavily on wood to meet energy requirements. According to [20], wood fuel share in Africa ranges 60–86% of primary energy consumption, 90–98% is consumed by household, and per capita wood fuel consumption is estimated at 0.89 m^3/year. Most of this wood is from forests and woodlands of hardwood wood tree species, and only a small volume is produced sustainably [17]. The use of wood for energy in developing countries continues to attract a great deal of attention, because the majority of the population in these countries face acute shortages of biomass energy due to the combined effects of increasing demand and diminishing supplies of this source of energy. As wood fuel becomes scarce, preferred tree species are harder to find or become locally extinct, and the women and children need to travel longer distances to reach supplies.

Presently, industrial wood in Zambia amounts to about 2.3 million m^3/annum, sawmilling accounts for 2 million m^3, while wood poles and other wood uses account for 0.3 million m^3.The wood industry is slowly growing scattered across the country comprising mainly sawmilling and pit-sawing units. The current consumption of sawn wood and all wood types (approx.. 600,000 m^3) is predicted to reach 980,000 m^3 by the year 2030. However, the sawmilling industry needs expansion of log resources and improvement of firms, recovery, value addition, skills, utilization of residues, and technology. There is only limited production of particle board, plywood, and block board produced in the country. The current consumption of wood-based panels (approx. 40,000 m^3) is predicted to attain about 110,000 m^3 by the year 2030 justifying the need to develop the wood-based panel industry. Wood furniture industry is low, but there is potential to improve it both at the firm level and sector level. The export opportunity exists for furniture to adjacent countries in Southern Africa. In addition, several investments including solar and bioenergy aiming to diversify the electricity generation in the country are mostly either in planning or implementing phase. According to the United Nations report [21], exports of wood from hardwood tree species has been increasing at an average annual change of 62% since 2010. This increase is probably due to the availability of markets and price premiums for industrial hardwoods in China and Europe. The increasing demand for hardwood tree species such as *Colophospermum mopane*, Kirk Ex J. Leornard, *Julbernardia paniculata* (Benth) Troupin, *Pterocarpus angolensis* DC, *Combretum molle* R.Br.ex G. Don, and other species belonging to Brachystegia species provide great opportunities for private sector involvement through value addition processing and bioenergy production [18]. However, raw material insecurity, unknown patterns of distribution, and uncertainty of the potential supply are often considered as potential risks to attracting investment in the wood industry in Zambia [18]. In 2005 and 2010, Zambia conducted a nationwide Integrated Land Use Assessment (ILUA) that provided data on various aspects of forests and woodlands to give information on the available growing stock for use by the wood industry sector, investors, policy-making, and other stakeholders [22]. However, the data require additional analysis targeting certain interests and other user groups. We hypothesized that wood fuel deficits vary by ecological regions and that Zambia's urban regions experience an energy crisis. The aim of this study was to assess the distribution, diversity, and abundance of the most suitable tree species for value-added renewable energy production from the miombo woodlands. This study's information is critical to developing strategies for the management of wood material for bioenergy production in developing countries in Africa and other parts of the world.

1.1 Domestic and commercial importance of selected hardwood tree species

The selected tropical hardwood tree species are preferred by most local people for both domestic and commercial use for a number of reasons to include color of wood, texture, density, and fiber for construction works [23]. In addition, the commercial importance and use of each of these species are influenced by economic factors such as availability, wood properties, and market acceptance [20]. The use of wood as a fuel for most local people plays an important role as an energy source. Brachystegia, Julbernardia, and Isoberlinia species are being the preferred genus for its high basic density and energy quality [24] for inclusion as biofuels in the form of firewood and its transformation into charcoal to obtain a more efficient fuel. According to [25], basic density of wood is a variable attribute because it changes according to geographical area and climate of the area. Moisture content is another important factor, and [26] reported that moisture contents of less than 8% are required to reduce the consumption of material to evaporate the water and also with these values a charcoal less susceptible to attack by biological agents is obtained. Twenty-five species studied have good potential as a source of locally available energy either in the form of firewood or charcoal with significantly better energetic properties. In addition, these species are abundant forest resources in the country, and hence their inclusion in management plans could be important in the development of the wood sector industry.

2. Materials and methods

2.1 Data acquisition

Inventory data were obtained from the Integrated Land Use Assessment inventory database maintained by Forestry Department, Ministry of Tourism, Environment and Natural Resources (MTENR) [27]. Although the ILUA II inventory covered the whole country, this study is based on filtered data from agro-ecological zones 1 and 2 covering Central, Eastern, Southern, and Western regions of the country (**Figure 1**). Zambia is broadly divided into three agro-ecological regions based on rainfall and soil and other climatic conditions [29]. These regions also experience high forest loss rates due to household heavy dependence on wood fuel (firewood and charcoal) collection to meet their energy requirements [30]. Earlier research studies also indicate that these regions are within a low rainfall projection for the future; hence, the area under miombo woodlands will decrease with up to 50% [31, 32], because poor rural farmers are heavily dependent on forest resources as safety nets.

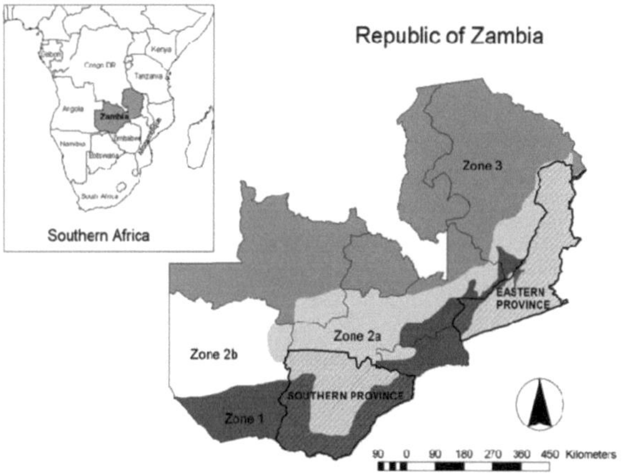

Figure 1.
Map of Zambia showing agro-ecological zones 1 and 2 covering parts of Eastern, Central, Southern, and Western provinces on which assessments for this study were based [28].

2.2 Rationale for selecting tree species

Tree species were initially selected following a survey. A team of experts conducted ethnobotanical survey through semi-structured interviews with selected community informants to identify the most preferred wood fuel tree species. These data were compared to the national database of the Integrated Land Use Assessment (ILUA) at the Forestry Department, Zambia [27]. A list of 25 tree species were drawn up from different families of which 15 of these species are listed in **Table 1**, and the most important included Fabaceae (Caesalpinioideae), Leguminosae (Mimosoideae), Combretaceae, and Mimosaceae (**Table 2**). The list of species was checked for spelling errors and scientific names, and those bearing different names were noticed and corrected accordingly. To obtain valid results in terms of tree species abundance and distribution, any unique names of species from the inventory data and duplications or spelling mistakes that would have affected the evaluation of abundance were removed. The rationale behind the selection of these species is that these are commonly used timber and wood fuel species and among the most preferred species of the miombo woodland [24, 33] on the basis of their commercial and economic importance. The studied species were further classified by their forest successional status based on literature review, expert consultation, and the authors' own field experience.

2.3 Data filtering and cleaning

The Integrated Land Use Assessment (ILUA II) raw data set collected within the limits of the tracts, plots, and subplots was acquired from the national database at Forestry Department, Ministry of Lands, Environment and Natural Resources, Lusaka, Zambia [27]. To determine the diversity, relative abundance, and distribution of the 25 most preferred wood fuel hardwood tree species for renewable energy, we filtered the data and grouped these species from agro-ecological zones 1 and 2 using Microsoft Excel. The filtered raw data were further cleaned. Data obtained include tree species scientific names, diameter, merchantable height, and GPS locations.

Species	Uses	Vegetation	Successional status
Brachystegia spiciformis	FE, CC	P,S	LS
Brachystegia floribunda	CC, FE	P,S	MS
Brachystegia longifolia	CC, FE	P,S	MS
Brachystegia boehmii	FE, CC	P,S	MS
Julbernardia globiflora	CC, FE	P,S	MS
Julbernardia paniculata	CC, FE	P,S	MS
Isoberlinia angolensis	FE, CC	P,S	MS
Colophospermum mopane	FE, CC	S	ES
Combretum molle	FE, CC	S	ES
Albizia versicolor	FE, CC	P,S	MS
Erythrophleum africanum	CC, FE	P,S	ES
Terminalia mollis	FE, CC	S	ES
Albizia antunesiana	FE, CC	P,S	MS
Pericopsis angolensis	FE, CC	P,S	ES
Acacia nigrescens	FE, CC	P,S	LS

Notes: CC = charcoal, FE = firewood, P = primary forest, S = secondary forest, ES = early successional, LS = late successional, MS = mid-successional.

Table 1.
Information on the most common 15 of the 25 selected tree species.

Family name	No. of species	% of family
Fabaceae	10	40
Caesalpinioideae	7	28
Combretaceae	3	12
Mimosaceae	1	4
Phyllanthaceae	1	4
Chrysobalanaceae	1	4
Papilionaceae	1	4
Apocynaceae	1	4
Total	25	100

Table 2.
List of families, number of species in each family, and percentage of each family of the tree species recorded.

3. Data analysis

The filtered data were organized and recorded in Microsoft Excel 2019 data sheet. Using the filtered and cleaned data from the national data set, we computed species densities, diameter distribution, abundance, and diversity for the wood fuel preferred hardwood tree species. All tree species were assigned to families, and relative diversity (number of species in a family) was obtained for tree species diversity classification.

3.1 Diameter class distribution

Does forest structure vary by region of the study area? In order to address this, density, diameter at breast height (DBH), and frequency were used for the description of vegetation structure. Diameter distribution is one of the most commonly and widely studied variables in wood science studies [34, 35] including tropical hardwood tree species [36]. Knowing the status of diameter distribution in the forest woodland provides information about forest structural attributes and serves as an integral part of wood management and planning for the near future. Density is the number of individuals per unit area or volume. The density of a species exhibits the abundance of a species in a given area [37]. The density was determined for each selected tree species from absolute density calculated from the total number of individual of a species counted in a plot divided by the total area sampled (0.1 ha). Relative density (RD) is the study of the numerical strength of a species in relation to the total number of individuals of all the species and was calculated as:

$$\text{Relative density (RD)} = \frac{\text{Number of individuals of tree species}}{\text{Total number of individuals}} \times 100 \tag{1}$$

Basal area for each tree species was calculated with the basal area function as follows:

$$\text{BA} = \frac{\pi d^2}{4(100)^2} \tag{2}$$

where BA is the basal area (m^2), π is the constant (3.142), and d is the diameter at breast height (cm). The total basal area of each species was obtained by adding together the basal areas of the individual trees of the species. Basal area of each species per hectare was estimated by extrapolating the total basal areas of the species using the formula as:

$$\text{BA} = \frac{H}{A} \times D \tag{3}$$

where BA is the basal area, H is the one hectare, A is the area of plot in hectares, and D is the basal area in each plot.

The frequency is a statistical parameter which reflects the spread of a species in a given area:

$$\text{Frequency} = \frac{\text{Number of plots in which a species occurs}}{\text{Total number of plots sampled}} \times 100 \tag{4}$$

The relative frequency of species was obtained from absolute frequency, dividing the number of sampling units in which the species occurs by the total number of sampling units as follows:

$$\text{Relative frequency (RF)} = \frac{\text{Frequency of respective tree species}}{\text{Total frequency of all tree species}} \times 100 \tag{5}$$

Regeneration status of the woodland was analyzed by comparing saplings and seedlings with mature trees of the area to insure adequate wood products supply. According to [27], the status was considered good regeneration if number of seedlings were greater than saplings and greater than mature trees; the regeneration status was relatively fair if number of seedlings were greater or less than or equal to saplings and less than or equal to mature trees; the status was poor regeneration if the species survives only under the first category.

3.2 Distribution and abundance of selected tree species

What influences the relative abundance of tree species for wood fuel? To answer this question, we computed the Shannon-Wiener index (H′) to measure species abundance and richness and quantify the tree species [38, 39]. Shannon-Wiener index is defined as:

$$H = -\sum_{i-1}^{s}\left(\frac{ni}{n}\right)x\ ln\left(\frac{ni}{n}\right) \quad (6)$$

where (H) is the Shannon-Wiener index, *ni* is the number of individuals of species *i*, *n* is the overall number of trees surveyed in the plot, *ln* is the natural logarithm, *S* is the number of species, and $\sum$ is the sum of the calculations. The index increases with the number of species in the area [2] and incorporates the species richness and the proportion of each species in all sampled plots (evenness) [40]. The Shannon diversity index ranges often from 1.5 to 3.5 and rarely reaches 4.5 [41]. The knowledge of species diversity is useful for establishing the influence of resource management, human disturbance, and the state of succession and stability in the area [42, 43]. Pielou's measure of species evenness formula was used to calculate species evenness defined as follows:

$$\mathrm{J} = \mathrm{H}'/\ln(\mathrm{S}) \quad (7)$$

where J is the evenness, H is the Shannon-Wiener diversity index, ln is the natural logarithm, and S is the number of tree species recorded in the considered plot [44, 45]. According to [46], a value for evenness is usually in the range 0 to 1, with 1 indicating that all species have the same abundance. Margalef's index was used as a simple measure of species richness [8]. Margalef's index is defined as:

$$(\mathrm{S}-1)/\ln \mathrm{N} \quad (8)$$

where S is the total number of species, N is the total number of individuals in the sample, and ln is the natural logarithm measurement of evenness for calculating species' evenness. The economic and ecological importance of tree species in agro-ecological regions 1 and 2 was analyzed using the Importance Value Index which is the summation of the relative values of frequency, abundance, and dominance [47] and is useful to compare the values of species [43]. According to [48], species with the highest importance value are the leading dominant species of specified vegetation. The Importance Vlue Index (IVI) describes the floristic structure and composition of forest woodlands and has often been used in the miombo forest systems [37, 49, 50]:

$$\text{Importance Value Index (IVI)} = \mathrm{RF} + \mathrm{RA} + \mathrm{RD} \quad (9)$$

where IVI is the Importance Value Index, RD the is relative dominance, RF is the relative frequency, and RA is the relative abundance.

3.3 Abundance and distribution of tree species

What tree species are preferred for each region? In order to address this question, we determined the abundance and distribution of wood fuel tree species for each region, and data were obtained from the national database of the Integrated Land Use Assessment (ILUA II) at Forestry Department, Zambia [27]. Based on expert knowledge and secondary information sources of preferred hardwood wood fuel tree species on high demand [18, 51], we filtered the commonly used wood fuel tree species from this database and the filtered data were cleaned to remove errors. Abundance (A) = Number of respective tree species/extent of sampling area. Using these data, we computed relative abundance based on the formula as follows:

$$\text{Relative abundance} = \frac{\text{Abundance of respective tree species}}{\text{Total abundance of all tree species}} \times 100 \qquad (10)$$

4. Results

4.1 Tree species abundance, distribution, and frequency

What influences the relative abundance of tree species for wood fuel? Abundance of wood fuel tree species in the study area may be influenced by many environmental factors; among these variables, soil type and topographic variables and other factors related to human impact are probably considered the most significant affecting species diversity and woody vegetation. However, this may require further investigations. The most frequently distributed tree species in the study area was *C. mopane* (6.5%) followed by *Julbernardia paniculata* (2.9%), *Brachystegia boehmii* (2.3%), *Brachystegia spiciformis* (1.8%), and the least was *Brachystegia globiflora* (1.2%). The frequency distribution of selected tree species in the study area is quite variable which can be explained by frequent use of these tree species in the study area for charcoal making and construction material. The most commonly used trees species for charcoal production are *Combretum molle, C. mopane, Isoberlinia angolensis, Brachystegia*, and Julbernardia species [42, 52]. The tree species with highest relative abundance was also dominant on the frequency of each individual tree. The higher the frequency, the higher the dominance index and vice versa. The relative abundance, relative density, basal area, frequency, relative dominance, and importance value indices of selected tree species for wood fuel energy production are shown in **Table 3**. The result shows the most abundant of these tree species being *C. mopane* (Benth), J. Leonard (550), with a relative abundance of 14.5%. It was followed by *Julbernardia paniculata* (Benth), Troupin (450), with a relative abundance of 11.9%, *Brachystegia boehmii* Taub. (396), with a relative abundance of 10.4%, *Brachystegia spiciformis* Benth (300) with relative abundance of 7.9%, *Diplorhynchus condylocarpon* Mull.Arg (337) with relative abundance of 8.9%, *Pseudolachnostylis maprouneifolia* Pax(270), with relative abundance of 7.2%, *Brachystegia longifolia* Benth. (218), with relative abundance of 5.8%, and *Combretum molle* R.Br.ex G.Don (195), with relative abundance of 5.2%, and the rest of species had relative abundance less than 5%. At the provincial level, the distribution of the hardwoods on demand was highest in the Western province with

No	Scientific name	No of individuals	BA (m^2)	F%	RA%	RD%	RF%	IVI%
1	Colophospermum mopane	550	22.4	6.5	14.5	21.4	21.5	57.4
2	Julbernardia paniculata	450	15.2	2.9	11.9	9.6	9.6	31.1
3	Brachystegia boehmii	396	13.1	2.3	10.4	7.6	7.6	25.6
4	Brachystegia spiciformis	300	13.7	1.8	7.9	5.9	5.9	19.7
5	Erythrophleum africanum	182	0.9	1.8	4.8	5.9	5.9	16.6
6	Diplorhynchus condylocarpon	337	5.6	1.6	8.9	5.3	5.3	19.5
7	Pseudolachnostylis maprouneifolia	270	0.8	1.3	7.2	4.3	4.3	15.8
8	Brachystegia utilis	75	2.9	1.3	2.0	4.3	4.3	10.6
9	Brachystegia longifolia	218	5.7	1.3	5.8	4.3	4.3	14.4
10	Brachystegia floribunda	123	3.4	1.2	3.5	4.0	4.0	11.5
11	Combretum molle	195	4.5	1.0	5.2	3.3	3.3	11.8
12	I. angolensis	52	2.0	0.7	1.4	2.3	2.3	6.0
Summary		**3148**	—	23.7	83.5	78.2	78.3	
Other species		**638**	—	6.6	16.5	21.8	21.7	

Note: BA = basal area, F = frequency, RA = relative abundance, RD = relative density, RF = relative frequency, IVI = Importance Value Index

Table 3.
Species, abundance, frequency, basal area, and Importance Value Indexes of most commonly used trees for fuel wood observed in agro-ecological zone.

relative abundance of 34.1%, followed by Luapula (19.7%) and Northern province (10.8%). The relative distribution of these hardwoods equally varied greatly across other provinces (i.e. Central 3.4%, Eastern 3.9%, and Southern 5.1%). Among the 200 hardwood species known for timber and energy found in Zambia, *P. angolensis* is known to be the natural resource among most top species in use [22]. Other species such as *Julbernardia paniculata*, *Brachystegia bohemii*, *C. mopane*, and several other Brachystegia species which were reported as the most abundant hardwoods in Zambia should be promoted for bioenergy, sustainable charcoal production, and carbon trade [53].

4.2 Diameter class distribution

Bar graphs were developed using the DBH versus number of individuals for seven arbitrary diameter classes (2.5–7.5 cm), (7.5–11.5 cm), (11.5–15.5 cm), (15.5–19.5 cm), (19.5–23.5 cm), (23.5–27.5 cm), and (≥27.5 cm) of the selected tree species (**Figure 2**).

Based on the assessment of diameter class distribution, the population structure pattern of the selected tree species recorded from the study area exhibited reverse J-shaped distribution (**Figure 2**). Inverse J-shaped pattern shows more trees in the small diameter classes (4-20 cm). The result showed significance at $p < 0.05$, t (24) = 2.73, and p = .05 despite small diameter trees (M = 99,SD = 89) recording largest number of trees than larger diameter class trees (M = 43, SD = 45), R-squared (R^2) = 90.7%, correlation (R) = 0.908, F statistic = 232.7), and p-value = 0.0087. This may be indicative of good reproduction and growing population in which young trees will

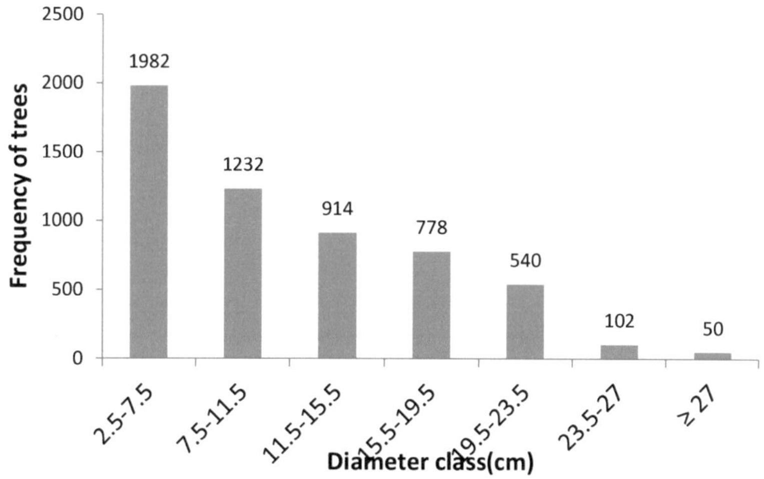

Figure 2.
Diameter distribution of standing trees ≥2.5 Dbh by diameter classes in agro-ecological zones 1 and 2.

grow into mature size classes. This may also suggest the occurrence of high disturbance of older trees due to frequent harvesting of trees for charcoal production, construction works, timber, and other miscellaneous uses. Tree frequency decreased with increasing diameter which is common for miombo woodlands with selective cutting where regeneration is active, an indication of a healthy recruitment of the individuals in the area [45, 50]. Other research studies within the miombo woodlands reported similar size class distribution [54, 55]. However, [56] argues that caution must be exercised in the use of inverse J-distribution as stock control in management, since the distribution assumes equal mortality rates among size classes and regarded it as biologically unrealistic.

4.3 Forest composition, structure, and richness

Do forest structure and composition for wood fuel tree species vary by region? Composition is used to refer to all plant species found in a stand or landscape including trees, shrubs, and grasses. Structure refers to the physical arrangement of various physical and biological components of an ecological system. In the study area, structure varied between agro-ecological zones 1 and 2 (**Figure 3**). In earlier research studies, it was reported that about 40% of Zambia's forest land is described as

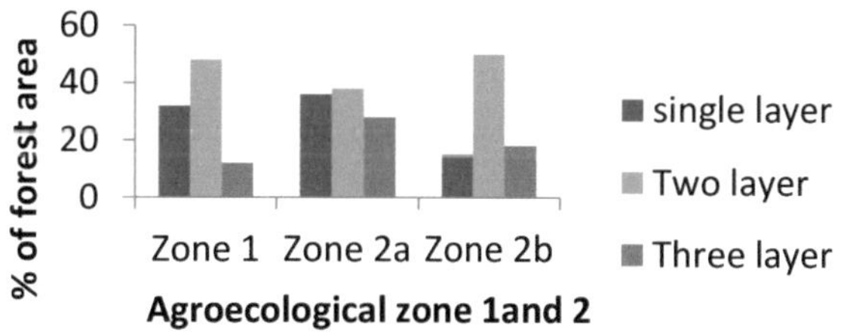

Figure 3.
Stand structure in forest land showing vegetation types (single-layer, two-layer, and three-layer) in agro-ecological zones 1 and 2 (Southern (zone 1), Central (zone 2a), and Eastern (zone 2b).

two-layer vegetation [22]; hence, the forest is characterized by two distinct canopy layers. In the study area, the forest with stand structure of single layer ranged 15–36%, two-layer vegetation ranged 38–50%, and three-layer vegetation ranged 12–30% (**Figure 3**). The highest proportion of forest land (≥30%) with single-layer vegetation was observed in Southern and Central provinces, suggesting that tree cutting for charcoal production and construction activities is high in these areas. A total of 6486 individuals with diameter ≥ 2.5 at breast height representing 149 tree species were recorded in agro-ecological zones 1 and 2 from which 25 species were selected for this study representing 6 families and 13 genera (**Table 2**). The species richness (149 species) of trees and the densities of species observed in this study compare well with miombo woodland occurring in other areas such as reported in Tanzania, Zimbabwe, and Zambia [2, 33, 37, 49, 57]. In Zambia, [58] enumerated a total of 238 woody species belonging to 154 genera showing high diversity and endemism over the country. The high number of species richness in the study area may be attributed to the pressure that plants undergo due to dry conditions in the two regions that contribute to emerging and growth of many new tree species. Climatic, edaphic variability, and anthropogenic activities are other factors associated with species richness [2]. According to [59], anthropogenic activities play a big role in the dynamics of miombo woodlands. The average tree density of 150 ± 30 trees per/ha is comparable with those reported by [2, 37]. Basal area provides an excellent indicator of the degree of stocking in a forest stand. This study showed that *Colophospermum mopane* (Benth), J.Leonard, gave a basal area of (22.4 m^2) per hectare followed by *Julbernardia paniculata* (Benth), Troupin, with the basal area of 15.2 m^2 suggesting that these tree species are well stocked, while *Acacia nigrescens* Oliv. (Knobthorn) with the basal area of 0.25 m^2 was understocked. The low stem density and basal area recorded in this study may be a consequence of frequent forest fires [27] and increased wood exploitation rate for charcoal production and construction materials [29, 42] in the area. Reference [58] observed a decline in biomass of some of the most productive and dominant species (*B. spiciformis, C.mopane, and I.angolensis*) in Zambia despite their high productivity. These species are heavily extracted by local people in their daily livelihood activities. The decline in biomass may be associated with low stem density and basal area recorded in this study.

4.4 Multipurpose values of selected tree species

Some of the selected tree species in the study area were found to have multipurpose values such as fuel wood (charcoal and firewood), timber, construction material, and other miscellaneous uses (**Table 1**). These species, *Julbernardia paniculata*, *Tamarindius indica*, *Pericopsis angolensis*, *Parinari curatellifolia*, and *Brachystegia spiciformis* were among the most commonly used trees for charcoal production and timber [16]. It is common that these species are rapidly removed from woodland once charcoal production starts in an area, hence affecting the species composition of miombo woodland [29]. The increase in demand for charcoal in Zambia and other parts of Africa has led to depletion of preferred hardwood tree species resulting in the use of a wider range of other tree species, suggesting that forests may be changing in structure and composition even in the near future. Where consumption is high, unselective harvesting has overexploited certain species especially near urban areas, and this suggests for the need to promote sustainable use of lesser known tree species for the wood industry [60].

4.5 Regeneration of tree species

The active regeneration and recruitment of tree species in agro-ecological regions 1 and 2 as depicted by this study is a good indication of sustainability of the forest woodlands which can ensure sustainable supply of wood products, hence sustaining the livelihoods of local people who depend on the forest woodland [61]. The middle diameter class (1–19 cm) had the highest number of stems, species, and family. Diameter for the tree species was inclined toward the small-sized diameter with diameter breast height below 20 cm. In the study area, different diameter distributions were observed, suggesting human disturbances which are probably linked with farming and tree harvesting activities and bush fires. A large proportion of total regeneration comprised small seedlings (70–80%) comparably to saplings (**Figure 4**). The highest density of seedlings was found in Eastern and Western provinces, while the lowest was recorded for Southern province due to frequent tree cutting for construction activities, fuel wood and charcoal production. The distribution of tree species such as *Julbernardia paniculata* and *Brachystegia boehmii* was widely spread in Central province, while *C. mopane* was concentrated in Eastern province. Disturbances in miombo woodlands to include uncontrolled harvesting of trees by local people for wood fuel and building materials are most likely to affect the size class distribution of the harvested trees and hence may change the composition and structure of tree species diversity and subsequently may result in increased tree mortality [56]. Efforts to enforce the forest law and regulations by government to provide absolute protection from illegal activities seem to be difficult because of the big demand of the wood products and services by local people for their livelihoods. However, to our understanding, there are no empirical studies in the study area to assess the impacts of anthropogenic disturbances on diameter distribution and forest health, and such information is required for sustainability management of these ecosystems.

4.6 Species density and basal area

C. mopane (CM) Kirk Ex J.Leonard (Fabaceae) had the highest number of stems (550 stems/ha) and a relative density of 21.4 and basal area of 22.4 m^2 (**Table 3**); hence, it was regarded as the most abundant species in the study area. This was followed by *Julbernardia paniculata (JP)* (Benth) Troupin (Fabaceae) with 450 stems/ha, relative density of 9.6 and basal area of 15.2 m^2. The third abundant species was *Brachystegia boehmii (BH)* (Taub) (Leguminosae) with 396 stems/ha, relative density of 7.6, and basal area of 13.1 m^2, *Diplorhynchus condylocarpon (DC)* (Welw ex Ficalho

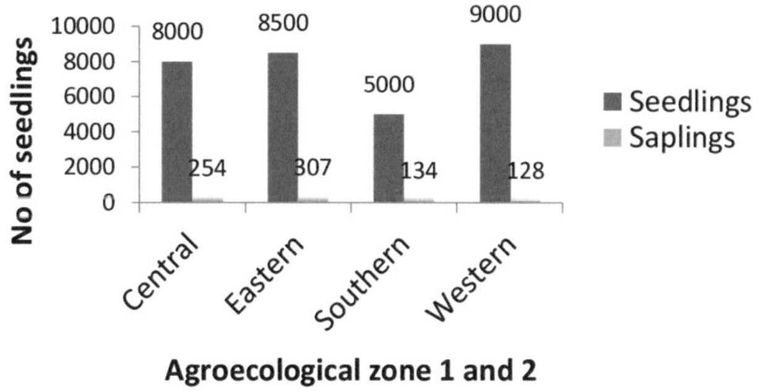

Figure 4.
Number of seedlings and saplings recorded regenerating in agro-ecological zones 1 and 2.

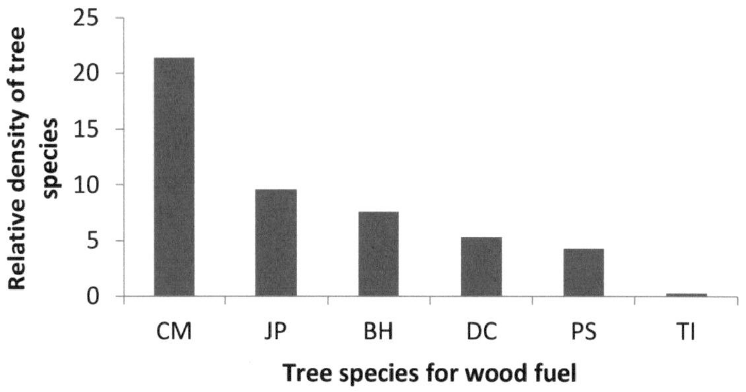

Figure 5.
Relative density of wood fuel tree species abundance in agro-ecological zones 1 and 2. Note: CM = Colophospermum mopane, JP = Julbernardia paniculata, BH = Brachystegia boehmii, DC = Diplorhynchus condylocarpon, PS = Pseudolachnostylis maprounefolia, TI = Tamarindus indica.

&Hiern) (Apocynaceae) with 337 stems/ha, relative density of 5.3, and basal area of 5.6 m^2, and *Tamarindus indica (TI)*, L (Fabaceae) lone tree assessed in the area had the largest mean DBH of 85 cm, relative density of 0.3, and basal area of 1.1 m^2, while the least DBH of 6 cm, with 270 stems/ha, relative density of 4.3, and basal area of 0.8 m^2 for the selected species was recorded for *Pseudolachynostylis maprouneifolia (PS)* Pax (Phyllanthaceae) (**Figure 5**).

4.7 Species diversity and Importance Value Index

What tree species are appropriate for each region? There is limited information on wood fuel species appropriate for each agro-ecological zone, particularly high altitude (≥2000 m) and semiarid zones. The abundance of wood fuel tree species in these areas is primarily based on rainfall, soil type, human activities, and also other climatic characteristics. Generally, the appropriate wood fuel tree species for each region are key legume trees of miombo and mopane woodlands, Brachystegia Benth (*Brachystegia boehmii* and *Brachystegia spiciformis*), *Julbernardia globiflora* (Benth) Troupin, *Pterocarpus angolensis* DC, and *Colospermum mopane.* Species diversity is a complex ecological characteristic of a particular forest area, that is, commonly used representation of ecological diversity. It contributes to ecosystem health and can be measured from the number of species (species richness) and relative abundance of individuals within each species (species abundance) [62]. Importance Value Index (IVI) gives knowledge on most important economically or ecologically tree species of a particular forest area. In this study based on Importance Value Index, 12 dominant thus economically and ecologically most important tree species (**Table 3**) are *Colophospermum mopane* (J.Kirk ex Benth), 57.4% was most dominant species followed by *Julbernardia paniculata* Benth. Troupin (31.1%), *Brachystegia boehmii* Taub. (25.6%), *Brachystegia spiciformis* Benth. (19.7%), Diplorhynchus condylocarpon Mull.Arg. (19.5%), *Erythrophleum africanum* (Welw.ex Benth) Harm (16.6%), *Pseudolachnostylis maprouneifolia* Pax. (15.8%), *Brachystegia longifolia* Benth. (14.4%), *Combretum molle* R.Br.ex G. Don (11.8%), *Brachystegia floribunda* Benth. (11.5%), *Combretum zeyheri* Sond. (11.1%), and *Brachystegia utilis* Hutch. (10.6%). The results show that the most important tree species in the study area have high diversity in the scale of Shannon-Weiner index of diversity. According to [63], Importance Value Index (IVI) ranks

Parameter	Values
Richness (total number of species)	146
Density (stems/ha)	150
Shannon-Wiener index (H′)	2.75
Pielou's evenness index(E)	0.85
Margalef's evenness index	2.9
Simpson's index of diversity	0.92

Table 4.
Characteristics of selected tree species in agro-ecological regions 1 and 2.

species in a way as to give an indication on which species come out as important element of the miombo tree species. The study revealed Shannon-Wiener index of diversity (H′) of 2.75, Pielou's index of 0.85, Simpson index of diversity of 0.92, and Margalef's Evenness index of 2.91 of the selected wood fuel tree species of the miombo woodland (**Table 4**). The Shannon index tells about species richness (number of species) and evenness (species distribution). Reference [39] reported that the larger the value of H′, the greater the species diversity and vice versa. An ecosystem with H′ value greater than 2 has been regarded as medium to high diverse [2]. This study shows higher diversity comparably to other studies in the miombo woodland where Shannon indexes ranged from 1.05 to 1.25 [64–66] but quite similar to other diversity study records such as by [44] who reported Shannon-Wiener index of 3.03 to 3.64. The H′ value of 2.75 from this study may be due to frequent bush fires and anthropogenic disturbances that opened the canopy and provided regeneration of light-demanding species such as *Uapaca* and *Albizia species* [49].

5. Discussion

Overexploitation of valuable tree species of miombo woodland in Zambia and other parts of Africa not only causes severe environmental impacts but also obstructs long-term use of wood fuel tree species for income generation, construction materials, charcoal, and fuel wood supply. Wood fuel from woodlands contribute to security of income and livelihoods of rural and peri-urban households. Presently, the availability of wood for wood fuel production and the quality of forest woodlands are not being sufficiently addressed by both traditional authorities and the Forestry Department; hence, filling this gap may demand overseeing harvesting, production, and charcoal trade activities. While wood fuel production is on the increase in Zambia and many other developing countries, management systems and use of efficient harvesting and production methods lag behind. Therefore, there is need for doubling extension and advisory services on wood fuel management and providing adequate research information to extension field officers.

The characteristic diameter distribution of wood fuel tree species in the study area may indicate the typical disturbance regime and subsequently the type of silvicultural system that should be used to achieve sustainable utilization for the wood industry. The inverse J-shaped curve observed in the study seems to agree with the findings by [49], since the miombo species are exposed to periodic annual fires that occur in the miombo vegetation [24]. The abundant regeneration in the study area may suggest that miombo wood fuel tree species are generally resilient and hence show good regeneration capacity, but this may demand targeted management of wood fuel sources. The overall stand structure (negative exponential curve) indicates that older trees probably provide minimal shading to the regeneration and saplings, and this promotes good growth of regeneration in larger numbers to insure sufficient wood resources for the future. Therefore, in order to improve regeneration in areas with closed canopy of larger trees,the canopy should be opened up by cutting mature trees for wood fuel and timber to allow penetration of light to the ground that would further facilitate regeneration growth. The large number of regeneration may also indicate utilization regime of tree species by local people who frequently cut and use Brachystegia species (*B. bussei, B. utilis*, and *B. spiciformis*) for construction material (poles and fiber). Reference [58] observed that increased wood extraction for wood fuel production and construction material in Zambia showed a negative balance for some of the most preferred productive and dominant species. This resulted into a decline in biomass for wood fuel trees species such as *B. spiciformis*, *C. mopane*, and *I. angolensis* suggesting that despite their high productivity, these species are heavily extracted from the woodlands a practice that could be unsustainable in the near future.

6. Conclusion and recommendation

The study revealed that the miombo woodlands in agro-ecological zones I and II in Zambia fairly have good species composition and distribution. Miombo woodlands play a significant role in sustaining the livelihood of many poor people in rural and urban areas in Southern Africa. However, the woodlands are under threat from evergrowing population in need of more wood, overharvesting of preferred hardwood tree species for wood fuel, poles, and timber. This calls for integrated approach for improving wood supply management and restoration and monitoring of woodlands to ensure sustainable wood energy. The status of preferred tree species density, abundance, composition, and distribution is at critical stage of threat with extinction as shown by low frequency of occurrence per hectare and hence needs quick action to tackle the problem sustainably.

The relative abundance, composition, and distribution varied across tree species. *C. mopane* (Benth) J. Leonard recorded the highest relative abundance (14.5) and basal area (22.4 m^2) per hectare among the 25 selected preferred hardwood species followed by *Julbernardia paniculata* (Benth) Troupin with abundance (11.9) and basal area (15.2 m^2). The diameter class distribution showed an inverse J-shaped trend with more trees in the small diameter classes. Different diameter distribution was observed, suggesting human disturbances probably linked with cutting activities for construction materials and wood fuel. However, further research should be geared toward the effects of anthropogenic disturbances and environmental factors (climate, soil, and topography) on tree growth, abundance, and species population changes of important tree species in the area as this will help proper management of wood resources in these ecosystems. Efforts are needed to promote collaboative management practices of wood resources that will ensure sustainable supply of wood products and services. Awareness and orientation should be given to local communities on existing forest laws and regulations on sustainable use of valuable tree species as well as the negative effects of illegal harvesting activities to enable them appreciate the role of trees to the wood industry in their environment. In rural areas and urban areas, local people still predominantly use fuel wood and wood charcoal for their cooking. In view of the foregoing, there will be need to promote investments in solar and bioenergy to diversify the electricity generation in the country. Furthermore, in recognition of the abundant wood resources and the growth of paper and paper board consumption in Southern Africa, there will be need to expand the pulp and paper industry in the country.

Acknowledgements

We would like to thank the Food and Agriculture Organization (FAO) and the Government of Finland for the logistics and financial support for data collection of Integrated Land use (ILUA) project in Zambia. We are thankful to Forestry Department for allowing us access the data used for the research from national data set. Members of the Science Research group and staff members in the School of Natural Resources, Copperbelt University, are also acknowledged for general support. Thanks go to an anonymous reviewer whose comments helped to improve this manuscript.

Declaration of competing interest

Obote Shakacite, Phillimon Ngandwe, Donald Chungu, and Vincent Nyirenda, the authors of the paper, declare that there is no conflict of interest regarding this publication.

Author details

Obote Shakacite[1], Phillimon Ngandwe[2]*, Vincent Nyirenda[3] and Donald Chungu[4]

1 Department of Plant and Environmental Sciences, School of Natural Resources, The Copperbelt University, Kitwe, Zambia

2 Department of Biomaterial Science and Technology, School of Natural Resources, The Copperbelt University, Kitwe, Zambia

3 Department of Zoology and Aquatic Sciences, School of Natural Resources, The Copperbelt University, Kitwe, Zambia

4 Directorate of Distance Education and Open Learning, The Copperbelt University, Kitwe, Zambia

*Address all correspondence to: pngandwe2015@gmail.com

References

[1] Sebukeera C, Muramira E, et al. Forests and woodlands. In: Africa Environment Outlook 2 Our Environment, Our wealth, Chapter 6. UNEP; 2006

[2] Giliba RA, Boon EK, Kayombo CJ, Musamba EB, Kashindye AM, Shayo PF1. Species composition, richness and diversity in miombo woodland of Bereku Forest Reserve.Tanzania. Journal of Biodiversity. 2010;**2**:1-7

[3] Djuikouo MNK, Doucet JL, Nguembou CK, Lewis LS, Sonk'e B. Diversity and aboveground biomass in three tropical forest types in the Dja biosphere reserve, Cameroon. African Journal of Ecology. 2010;**48**:1053-1063

[4] Frost P. The ecology of miombo woodlands. In: Campbell B, editor. The Miombo in Transition: Woodlands and Welfare in Africa. Bogor, Indonesia: Center for International Forestry Research; 1996. pp. 11-58

[5] Ngoma J, Speer JH, Vinya R, Kruijt B, Moors E, Leemans R. The dendrochronological potential of Baikiaea plurijuga in Zambia. Dendrochronologia. 2017;**41**:65-77

[6] Massuque JZ, de Assis MR, Trugilho PF. Characterization of Miombo species used by rural communities as fuelwood in northern Mozambique. In: Energy Sources, Part A. Recovery, Utilization, and Environmental Effects. Southern Forests, Journal of Forest Science. 2020;**68**:520-528

[7] Cuvilas C, Lhate I, Jirjis R, Terziev N. The characterization of wood species from Mozambique as a fuel. Energy Sources, Part A. Recovery, Utilization, and Environmental Effects. 2014;**36**: 851-857

[8] Margalef R. Temporal succession and spatial heterogeneity in phytoplankton. In: Buzzati-Traverso, editor. Perspectives in Marine Biology. University of California Press Berkeley; 1958. pp. 323-347

[9] Desta HM, Ambaye SC. Determination of energy properties of fuel wood from five selected tree species of tropical highlands of Southeast Ethiopia. Journal of Energy. 2020;**363**:1-7

[10] N'gandwe P, Chungu D, Shakacite O, Vessa L. Abundance and distribution of top five most valuable Hardwood Timber Species in Zambia and their implications as sustainable supply. In: Mottonen V, Heinonen E, editors. 6th International Scientific Conference on Hardwood Processing Proceedings. Helnsiki, Finland: Natural Resources Institute; 2017

[11] CAMCO. Biomass Energy Strategy (BEST). Eschborn, Germany: Africa-EU Renewable Energy Cooperation Programme; 2014

[12] Clarke J, Cavendish M, Coote C. Rural households and miombo woodlands: Use, value and management. In: The Miombo in Transition: Woodlands and Welfare in Africa. Bogor, Indonesia: CIFOR; 1996

[13] Syampungani S, Tigabu M, Matakala N, Handavu F, Oden PC. Coppicing ability of dry miombo woodland species harvested for traditional charcoal production in Zambia: A win–win strategy for sustaining rural livelihoods and recovering a woodland ecosystem. Journal of Forest Research. 2016;**28**: 549-556

[14] Mugo F, Ong C. Lessons from eastern Africa's unsustainable charcoal business. Working paper, World

Agroforestry Center (ICRAF), Eastern African Regional Program, 2006.

[15] Phiri JS, Moonga E, Mwangase O, Chipeta G. Adaptation of Zambian agriculture to climate change-A comprehensive review of the utilization of the Agro-ecological regions. Zambia Academy of Science Policy Note. 2013; **27**:1-41

[16] Falca'o MP. Charcoal Production and Use in Mozambique, Malawi, Tanzania and Zambia: Historical Overview, Present Situation and Outlook. Maputo, Mozambique: International Network for Bamboo and Rattan (INBAR), Conference; 2008. pp. 1-20

[17] UNEP United Nations Environment Programme. Annual Report, In Review. 2019

[18] Gumbo DG, Moombe KB, Kandulu MM, Kabwe G, Ojanen M, Ndhovu E, et al. Dynamics of the Charcoal and Indigenous Timber Trade in Zambia: A Scoping Study in Eastern. Bongor, Indonesia: Center for International Forest Research (CIFOR); 2013;occasional paper **86**:1-69

[19] Alnatheer O. The potential contribution of renewable energy to electricity supply in Saudi Arabia. Energy Policy. 2005;**33**:2298-2312

[20] FAO. State of the World's Forests 2016: Forests and Agriculture: Land Use Challenges and Opportunities. Rome: FAO; 2016

[21] UN United Nations Comtrade Database. 2003. Available from: http://comtrade.un.org/ [Accessed April 10, 2022]

[22] Shakacite O, Chungu D, et al. Integrated Land Use Assessment Phase II – Report for Zambia. Lusaka, Zambia: The Food and Agriculture Organization of the United Nations and the Forestry Department, Ministry of Lands and Natural Resources; 2016

[23] Brinberg D, Bumgardner M, Daniloski K. Understanding perception of wood household furniture: Application of a policy capturing approach. Forest Products Journal. 2007; **57**(7/8):21-26

[24] Chidumayo EN. Miombo Ecology and Management. An Introduction. Sweden: Stockholm Environment Institute; 1997. pp. 22-30

[25] Ordonez-Diaz JAB, Galicia-Naranjo A, Venegas-Mancera NJ, Hernandez-Tejeda T, Ordonez-Diaz MJ, Davalos-Sotelo R. Densidad de las maderas mexicanas por tipo de vegetacion con base en la clasificacion de J. Rzedowski: Compilacion. Madera y Bosques. 2015; **21**:77-126

[26] Heya MN, Pournavad FR, Carrillo-Parra A, Colin-Urieta S. Bioenergy potential of shrub from native species of North-Eastern Mexico. International Journal of Agricultural Policy and Research. 2014;**2**:475-483

[27] Forestry Department. Ministry of Tourism, Environment and Natural Resources, Project Plan. Support to "Phase II of Integrated Land Use Assessment (ILUA) in the Republic of Zambia, Strengthening Forest Resources Management and Enhancing its Contribution to Sustainable Development, Land Use and Livelihoods," Lusaka, Zambia 2010

[28] Madder M, Mtambo J, Chaka G. Further evidence for geographic differentiation in R-appendiculatus (Acariodidae) from Wastern and Southern provinces of Zambia. Experimental and Applied Acarology. 2007;**41**:129-138

[29] Syampungani S, Chirwa PW, Akinnifesi FK, Ajayi OC. The miombo woodlands at crossroads: Potential threats, sustainable livelihoods, policy gaps and challenges. Natural Resources Forum. 2009;**33**:150-159

[30] Bwalya SM. Household dependence on Forest income in rural Zambia. Zambia Social Science Journal. 2013;**2**:1

[31] GEF. Climate Change Adaptation in Forest and Agricultural Mosaic Landscapes Project. Lusaka, Zambia: Global Environmental Facility (GEF), Ministry of Finance and National Planning; 2019

[32] UNISDR United Nations Office for disaster risk reduction, Annual Report 2018

[33] Luoga EJ. The Effect of Human Disturbances on Diversity and Dynamics of Eastern Tanzania Miombo Arborescent Species. Johannesburg: University of Witwaterrand; 2000

[34] Leak WB. Origin of sigmoid diameter distribution. United States Department of Agriculture, Forest Service, Northeastern Research Station. Research Paper. 2002;**718**:1-14

[35] Massuque J, Reis M, Breno DA, Massuque J, de Assis MR. Influence of lignin on wood carbonization and charcoal properties of miombo woodland native species. European Journal of Wood and Wood Products. 2021;**79**:527-535

[36] De Lima RB, Bufalino L, Junior A, da Silva JA, Ferreira RL. Diameter distribution in a Brazilian tropical dry forest domain: Predictions for the stand and species. Anais da Academia Brasileira de Ciencias. 2017;**89**:1189-1203

[37] Gonçalves FMM, Revermann R, Gomes AL, Aidar MPM, Finckh M, Juergens N. Tree species diversity and composition of Miombo woodlands in South-Central Angola: A chronosequence of forest recovery after shifting cultivation. International Journal of Forestry Research. 2017;**2017**:13

[38] Begon M, Townsend CRH, John L, Colin RT, John LH. Ecology: From Individuals to Ecosystems. Australia: Blackwell Publishers Hoboken; 2006

[39] Magurran AE. Measuring Biological Diversity. Oxford: Blackwells; 2004

[40] Heip CHR, Herman PMJ, Soetaert K. Indices of diversity and evenness. Oceanis. 1998;**24**:1-28

[41] Gainess LW, Richy JH, John FL. Monitoring Biodiversity: Quantification and Interpretation. Portland, Oregon, USA: Department of Agriculture, Forest Services, Pacific Northwest Research Station; 1999

[42] Monela GC, Abdallah JM. Principle socio-economic issues in utilization of Miombo Woodlands in Tanzania. In: Paper Presented in the Conference on Management of Indigenous Tree Species for Ecosystem Restoration and Wood Production in Semi-Arid Miombo Woodlands in Eastern Africa, Tanzania. Finnish Forest Research Institute. 2007; **50**:115-122

[43] Lamprecht H. Silviculture in the tropics. Tropical forest ecosystems and their tree species-possibilities and methods for their long-term utilization, Germany. Nature English. Eschborn, Germany: Deutsche Gesellschaft Fur Technische Zusammernarbit (GTZ) GmbH, TZ Verlag; 1989;**296**:1-296

[44] Hofico NSA, Fleig FD. Diversity and structure of Miombo woodlands in Mozambique using a range of sampling sizes. Journal of Agricultural Sciences and Technology. 2015;**5**:679-690

[45] Kent M. Vegetation Description and Analysis: A Practical Approach. 2nd ed. New York: Wiley-Blackwell; 2012

[46] Bonoua W, Glèlè Kaka R, Assogbadjo AE, Fonton HN, Sinsin B. Characterization of Afzelia africana Sm. Habitat in the Lama Forest Reserve of Benin. Forest Ecology and Management. 2009;**258**:1084-1092

[47] Gebrewahid Y. Biodiversity conservation through indigenous agricultural practices: Woody species composition, density and diversity along an altitudinal gradient of Northern Ethiopia. Cogent Food & Agriculture;**5** (1):1700744

[48] Shiburu S, Balcha G. Composition structure and regeneration status of woody plant species in didin natural forest, south East Ethiopia: An implication for conservation. Ethiopian Journal of Biological Sciences. 2004;**3**: 5-55

[49] Kalaba FK, Quinn CH, Dougill AJ, Vinya R. Floristic composition, species diversity and carbon storage in charcoal and agriculture fallows and management implications in miombo woodlands of Zambia. Forest Ecology and Management. 2013;**304**:99-109

[50] Jew EKK, Dougill AJ, Sallu SM, O'Connell J, Benton TG. Miombo woodland under threat: Consequences for tree diversity and carbon storage. Forest Ecology and Management. 2016; **361**:144-153

[51] Azanzi P, Putzel I, Gumbo D, Mupeta M. Rural livelihoods and the Chinese timber trade in Zambia's Western Province. International Forestry Review. 2014;**16**:447

[52] Syampungani S, Malambo FM. Opportunities and challenges for sustainable management of miombo woodlands: The Zambia perspective. Forest Research. 2008;**98**:125-130

[53] Hagglom and Partners. Zambia Forest Opportunities for Finnish Stakeholders. Finland: Ministry of Foreign Affairs; 2016

[54] Kalaba FK, Quinn CH, Andrew J, Dougill AJ. Carbon Storage, Biodiversity and Species Composition of miombo Woodlands in Recovery Trajectory after Charcoal Production and Slash and Burn Agriculture in Copperbelt Province in Zambia. Leeds: University of Leeds; 2012

[55] Zimudzi C, Mapaura A, Chapano C, Duri W. Woody species composition, structure and diversity of Mazowe botanical reserve. Journal Biodiversity and Environmental Sciences. 2013;**3**: 17-29

[56] Isango J. Stand structure and tree species composition of Tanzania Miombo woodlands: A case study from miombo woodlands of community based Forest Management in Iringa District. Forest Research. 2007;**50**:43-56

[57] Backéus I, Pettersson B, Strömquist L, Ruffo C. Tree communities and structural dynamics in miombo (Brachystegia-Julbernardia) Woodland, Tanzania. Forest Ecology and Management. 2007;**230**:171-178

[58] Pelletier J, Paquette A, Mbindo K, Zimba N, Siampale A, Chendauka B, et al. Carbon sink despite large deforestation in African tropical dry forests(miombo woodlands). Environmental Research Letters. 2018; **13**:094017

[59] Chidumayo EN. Land use, deforestation and reforestation in the Zambian Copperbelt. Land Degradation and Rehabilitation. 1989a;**1**:209-216

[60] Chidumayo EN, Marunda C. Dry forests and woodlands in sub-Saharan Africa: Context and challenges. In: Chidumayo EN, Gumbo DJ, editors. The Dry Forests and Woodlands of Africa: Managing for Products and Services. London: Earthscan Ltd; 2010. pp. 1-9

[61] Dk L, Maranga EK, Aboud AA, Cheboiwo JK. Role of Forest resources to local livelihoods: The case of east Mau Forest ecosystem, Kenya. International Journal of Forestry Research. 2016;**10**: 1-10

[62] Halmilton AJ. Species diversity or biodiversity? Journal of Environmental Management. 2005;**2019**(75):89-92

[63] Munishi PKT, Philipina F, Temu RPC, Pima NE. Tree species composition and local use in agricultural landscapes of west Usambaras, Tanzania. African Journal of Ecology. 2008;**46**:66-73

[64] Shirima DD, Munishi PKT, Lewis SL, Burgess ND, Marshall AR, Swetnam RD, et al. Carbon storage, structure and composition of miombo woodlands in Tanzania's eastern arc mountains. African Journal of Ecology. 2011; **2011**(49):332-342

[65] Williams M, Ryan CM, Rees RM, Sarnbane E, FernandoJ GJ. Carbon sequestration and biodiversity of re-growing miombo woodlands in Mozambique. Forest Ecology and Management. 2008;**254**:145-155

[66] ZDA Zambia Development Agency Annual Report 2011

Chapter 7

Management Challenges of Gambella National Park

Gatluak Gatkoth Rolkier and Mulugeta Ruot Kuon

Abstract

Gambella National Park has several management challenges that included previous settlement and current settlement and agricultural investment. The objectives of this study were to examine previous and current settlement and agricultural investment in the surrounding area of Gambella National Park. The method that was employed for this study was systematic sampling. ArcGIS version 10.5 was used for analyzing the data. Among seven Refugees camps, Pugnido 1 with a total population of 68,176 (17%) and Pugnido 2 with a population of 17,793 (14%) are close to Gambella National Park followed by Tirikedi 72,876 (18%). It also shows that some of the current resettlement or villages were wrong placed inside the Gambella National Park. Therefore, the current resettlement program becomes another challenge for management of Gambella National Park. The Karaturi, Rushi, and Saudi leased land are around vicinity and some part of Gambella National Park. Karaturi leased land includes 50,000 ha of land from the National Park. As a result of current agricultural investment, the previous National Park which was 5,06 km^2 is now reduced to 4,575 km^2. The institutional change indicated the serious turnover of staff at Gambella National Park.

Keywords: Gambella National Park, Settlements, Agricultural investment, management challenges, biodiversity

1. Introduction

There are many views regarding the management of protected areas [1]. ReedNoss, one of the most powerful conservation biologists in North America, claims that though management of national parks is a system of control, environmental management is however necessary in many areas in order to preserve its biological diversity, particularly when the area is imposed by a variety of disturbance regimes. Rick Searle, a former park naturalist with Parks Canada, however, advocates any mixture of approaches that ensures the most restoration and maintenance of wildness is viable [2]. Alternatively, the management of national parks can also be viewed as being fundamentally the management of people, such as the introduction of visitor shares to manage the inflow of people [3].

National Parks and other protected areas are today's main method in protecting and freeing areas from the pressures of development and as an attempt to safeguard its natural, cultural, or historic heritage. The problem of effective park management is becoming increasingly obvious due to external and internal burdens. Areas nearby national parks for instance are becoming ever more urbanized or exploited for industrial determinations. Internal pressures include the need for tourism to raise revenues and its following development requirements within the park boundary. These burdens combined make national park management a truthful challenge of today [4]. Struggles between people and Wildlife Conservation Authorities have increased around protected areas because of increasing human populations and their activities, such as settlements, agriculture, livestock husbandry, and other human activities. These conflicts include blockage of migratory and dispersal areas, loss of habitat for wildlife, raiding of crops, and attacks on livestock by wildlife. The other conflicts were the competition for resources such as watering points and grazing areas, illegal poaching, and disease transmission [5]. These activities are the challenges for management of National Parks.

Deforestation in and around protected areas is causing habitat loss and fragmentation which in turn impact the wild animals in many ways. Wild animals lose their habitat, sources of food, and spaces due to loss of biodiversity, land degradation, and encroachment [6]. Illegal hunting is a threat to wildlife conservation works in Africa and it is a common practice in Ethiopia. It can be fueled by various factors that can range from simple social prestige which is connected to bravery and manhood, to the need for wild animals' meat for consumption, and commercial trading of valuable parts of the animals [7].

2. Problem statement

Gambella National Park is very rich in term of its biodiversity; however, the management of the National Park facing so many challenges, and therefore, its biodiversity is under serious threat to human intervention. The human interventions were resettlement programs during previous regime and settlement of South Sudanese refugees in the adjacent areas of Park. The current human interventions are the need for development such as agricultural investment, villagilization program, and current civil war in South Sudan and poaching. As result, the management of wildlife resources in the park becomes great challenge and therefore, those challenges need to be studied so as to suggest better solutions for better management of the park.

3. Objectives

3.1 General objective

The general objective of this study is to investigate the challenges facing the management of Gambella National Park.

3.2 Specific objectives

The specific objectives are:

To examine the previous settlement program in the adjacent areas of the park.

To identify current resettlements in the surrounding areas of the park.

To investigate the current agricultural investment in the surrounding area of the park.

4. Materials and methods

4.1 Description of study area

Gambella National Park is located in the central part, mainly in the lowland plain of the Gambella Regional State. According to EWCO [8], the park is situated within latitude of 8°N and longitude of 34°15′E. And it is between nine administrative woredas namely Jikawo, Lare, Wantawo, Akobo, Gambella, Itang, Gog, Abobo, and Jor. It was established in 1973 with redemarcated area covered of 4,575 km^2 (**Figure 1**).

4.2 Method of data collection

4.2.1 Sampling design and data collection

There are eight investment board members of institutions in Gambella Regional State. This includes Investment Agency, Mine and Energy Agency, Bureau of Culture and Tourism, Bureau of Environment, Forest and Climate change, Gambella National Park office, Bureau of Agriculture and Natural Resources, Bureau of Trade and Industry, and Rural Road and Transport Agency. Accordingly, five experts from four institutions (Gambella National Park office, Investment Agency, Bureau Environment, Forest and Climate Change, and Bureau Agriculture and Natural Resources) were systematically selected for this study which makes up a total of 20 respondents from the region. Moreover, four representatives from selected woredas

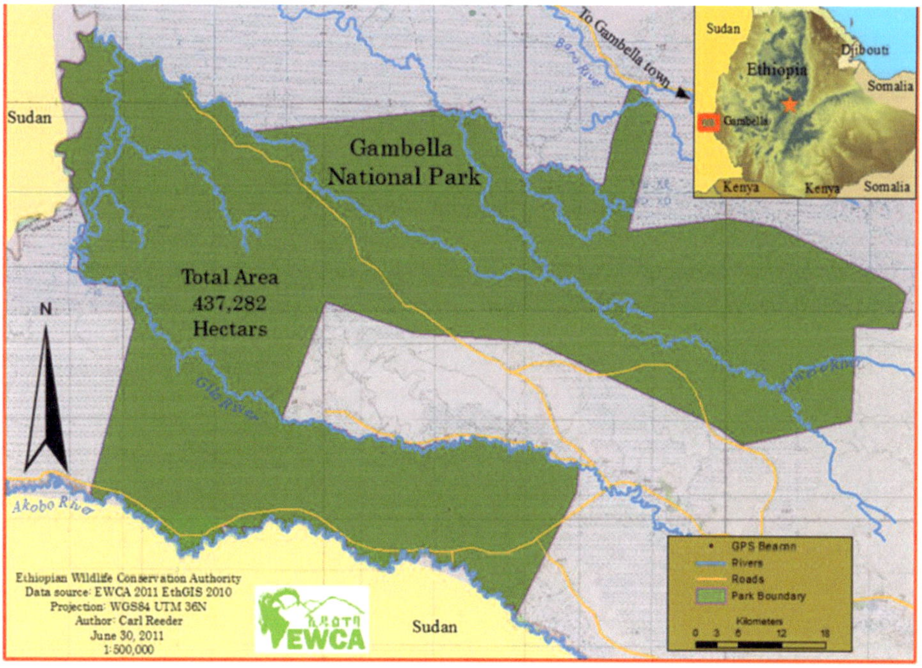

Figure 1.
Location of Gambella National Park within Gambella Regional State.

(Abobo, Itang special, and Wantaw) were also systematically selected, and 1 each from selected institutions (Agriculture and Natural Resources Offices, Administrative Offices, Culture and Tourism, and Rural Road Offices) which make up a total of 12 respondents. In Gambella National Park, five representatives from game rangers from three camps sites (Pokede, Corridor, and Mun) were systematically selected for this study which make up a total 15 of respondents. Therefore, there were total of 47 respondents. The field observation was also made in the main agricultural farms of target woreda adjacent to the national park.

4.2.2 Sources of data

The data collections for this study were collected from both secondary and primary sources. The secondary sources were mainly from unpublished and published sources. The unpublished sources included annual reports of selected Institutions. Literature review of other documents related to the challenges of Park management of biodiversity from different Libraries and Internets were also made to complement and refine the information that has been collected.

The primary sources were collected from the current agricultural investment and resettlements in the region and areas adjacent to National Park. These were collected from the experts working in the government institutions and field observations.

4.3 Data analysis

The collected data were summarized and analyzed by using ArcGIS 10.5 software taking each settlement and commercial agriculture as unit of analysis. Descriptive statistics such as mean, percentages, figures, tables, and standard error were calculated to present the results.

5. Results and discussion

5.1 Challenges facing management of biodiversity of the park

5.1.1 Impact of south sudanese refugees and SPLA on biodiversity of the park

There was increasing deforestation in the areas around all refugee camps where the refugees cleared the woodland for house construction and firewood. Hunting and poaching for food also decreased wild animals. According to the elderly informants some species of fish disappeared from fishing lakes and ponds such as Lake Tata and "Duma" or 'Kongdokuach' swamp because of increasing fishing by refugees. Because of the long duration of conflict in Southern Sudan, a large number of refugees had a profound impact on the local wildlife. After the arrival of the SPLA soldiers in the region in 1983 in general and their headquarter known as Bilpam in the Park in particular, they hunted wild animals such as buffalo, elephants, giraffe, white-eared kob, and Nile lechwe with automatic weapons for their own food. The traditional method of hunting by local community changed to hunting with automatic weapons they purchased from the SPLA soldiers. The local population had got army ammunition from the SPLA and therefore, both the local communities of Gambella and SPLA soldiers were heavily engaged in poaching.

According to current official information obtained from United Nations High Commissioner for Refugees (UNHCR), Gambella sub-office, there are seven camps in four woredas of the region. The three camps (Kule 54,547 (14%), Tirikedi 72,876 (18%), and Nguyiel 74,095 (18%)) are located at Itang special woreda. The two camps (Pugnido 1 68,176 (17%) and Pugnido 2 17,793 (14%)) are located at Gog woreda, whereas Jawie 62,641 (16%) and Okugo 13,016 (3%) are located at Gambella Zuria and Dimma woreda, respectively; **Figure 2**. Among Refugee's Camps, Pugnido 1 and 2 are close to Gambella National Park followed by Tirikedi. Therefore, the impacts on wildlife habitats such as cutting down of trees for firewood, house construction, and poaching were more observed at vicinity of those camps (Pugnido 1&2 and Tirkedi). Hence, care and maintenance need to be carryout by Gambella National Park Office, Administration Returnees Affair (ARRA), and UNHCR.

5.1.2 Impact of resettlement program on Wildlife resources in the park

Key informants stated that the 1984–1985 severe drought and famine in the northern part of the Ethiopia became a pretext for the Derg regime to carry out forced resettlement villages which were set up at Abobo, Gog, and Gambella woreda in adjacent part of Gambella National Park. About 60,000 highlanders were settled in the region. This brought adverse impact on vegetation and wild animals in the Gambella National Park because some of the resettled villages were placed inside the National park.

They also said that the current resettlement program of the region had also negative impacts on wildlife resources of the park. According to the information obtained

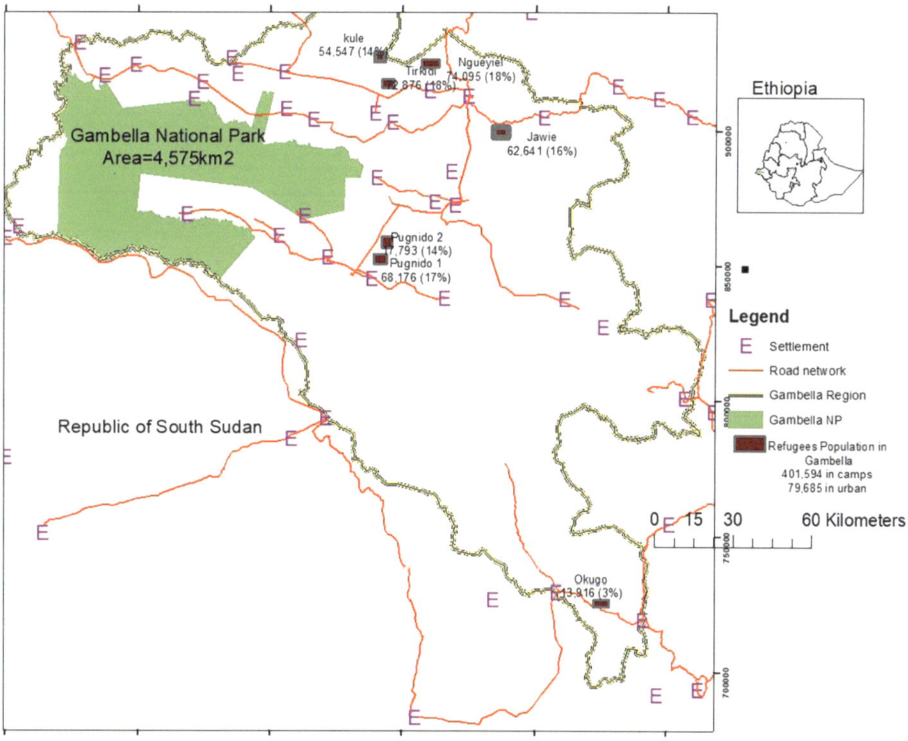

Figure 2.
Location of Refugee camps and their population within Gambella Regional State.

from Bureau of Agriculture of Gambella regional state, about 45,000 households were relocated from their original villages to cluster village. Some villages were not properly planned and wrongly placed inside the National Park, for instance, the villages named Ongoke at Jor woreda, Gire at Makuey woreda, and Kankan at Akobo woreda are placed inside the National Park. However, the Shentaw of Jor woreda is placed at adjacent area of the Park (**Figure 3**). Hence, the communities of these villages used to clear grasses and the trees for construction of their houses. Moreover, they use to hunt animal like buffalo and white-eared kob. Therefore, the current resettlement program becomes another challenge for management of the park.

5.1.3 Impact of current agricultural investment on park management

Agricultural investment statistic obtained from regional investment agency has shown that the current agricultural investments had impacted and decreased the boundary of Gambella National Park. About 100,000 ha of land from Gambella National Park was leased to both local and foreign investors. Karaturi Global of India, Saudi Star Agriculture Development Private limited company of Saudi Arabia, and Rushi have leased 100,000, 11,000 ha, and 25,000 ha of land in Gambella region respectively (**Figure 3**). According to the land administration and land use Department of Bureau of Agriculture of Gambella region, over 700 domestic investors are also engaged in agricultural production in Gambella region. Both the Karaturi, Rushi, and Saudi leased land are around vicinity and some part of the park areas. Karaturi leased land had included 50,000 ha of land from Gambella National

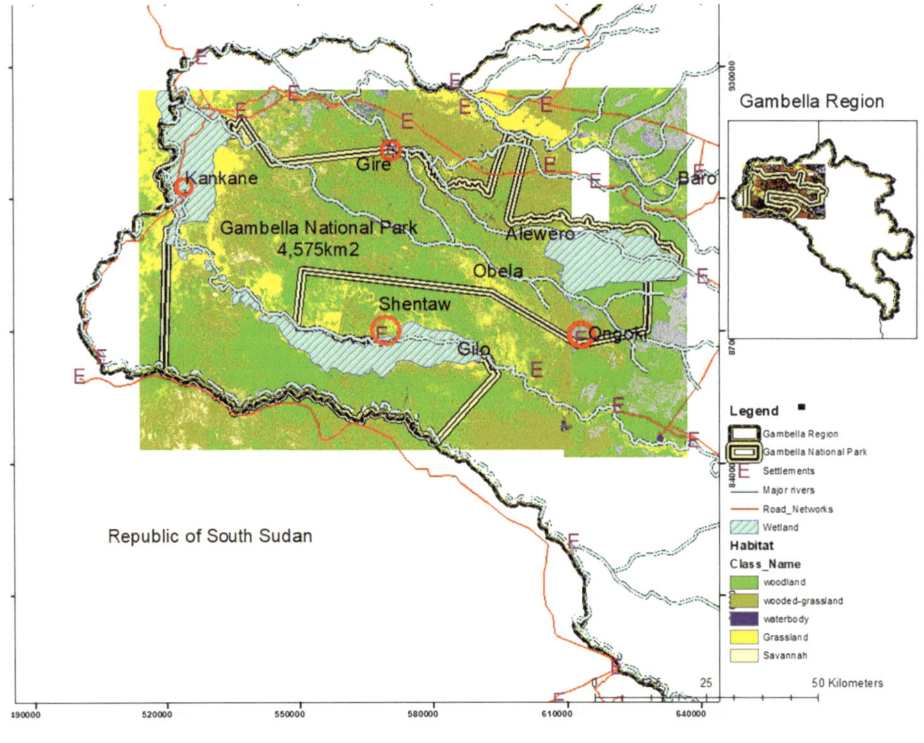

Figure 3.
Impact of settlement at Gambella National Park. Source: Gatluak's own data 2019.

park while the other leased land included 50,000 ha which was seen outside the current National Park leased by remaining investors. As a result of current agricultural investment, the previous National Park which was 5061 km^2 is now reduced to 4,575 km^2. Moreover, information obtained from key informants, the wetland locally known as "Duma" in Anuak and "Kongdokuach" in Nuer is being irrigated in its upper course by Saudi Star Company and irrigated from its lower course by Karaturi Company. This has a large impact on wildlife resources for both large mammals, such as Buffalo, Nile lechwe, and rare bird species like a shoe-bill stork and other aquatic animals such as fish species (**Figure 4**).

5.1.4 The management challenges associated with Institutional changes and its culture

Since the time of its establishment up to the first half of 1996, the management of the park has been run by Ethiopian Wildlife Conservation Organization (EWCO). During the decentralization policy of the government of Ethiopia, the responsibility of administering the park was handed over to the regional bureau of agriculture. In 2005/2006, the warden of Gambella National Park was transformed by regional government into as political post. Furthermore, during the regional structure in 2007, the park office has been combined with culture and tourism bureau. At the end of 2008, the park office was combined with bureau of agriculture of the region. In late 2009, the administration of the park has returned to Ethiopian Wildlife Conservation

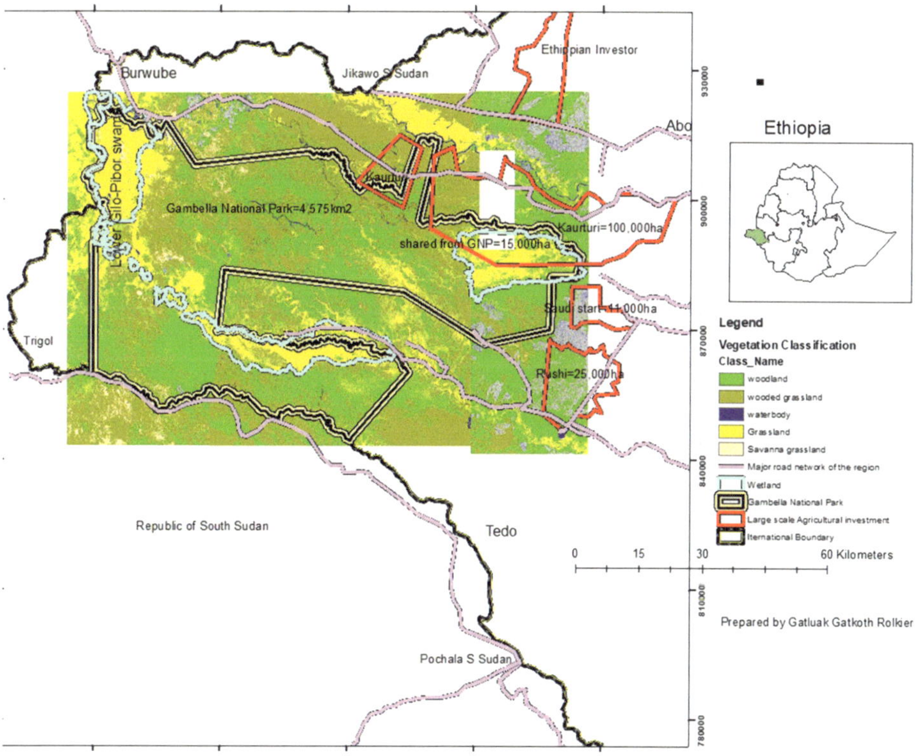

Figure 4.
Agricultural investment in adjacent to park within Gambella Regional State. Gatluak's own data 2019.

No	Post	No	Status	Salary (Birr)	Remark
1	Warden (Office Manager)	1	Not employed	**7647**	Leaved 2017
2	Secretary	1	Present at work	2,500	
3	Wildlife Senior expert	1	**Not employed**	6992	
4	Wildlife junior expert	1	Present at work	**2,800**	
5	Wildlife Regulatory expert	1	**Not employed**	6992	
6	Tourism beginner expert	1	Present at work	3526	
7	Community Services expert	1	Present at work	4458	
8	Animal health assistant	1	**Not employed**	3526	
9	Head of Scouts	8	Present at work	4439	
10	Scout	103	Only 36 are employed	2244	
11	Supporting Staff services	15	Seven are present at work and **eight are not employed**	2,500–4,458	
Total		134			

Table 1.
Gambella National Park Staff Composition and indication of staff turnover based their income (Salary).

Authority. The problems here associated with this institutional change were serious turnover in staff and lack of proper care for official documents. Other institutional problems mentioned by Gambella National Park experts were lack of transparency from both the Wildlife Conservation Authority and Gambella National Park, small amount of salary of staff as compared to Gambella Regional State staff salary, lack of motivation and so many positions without employees. For instance, Chief Warden, Procurement and Finance Officer, revenue expert and recorded and documentation expert are all carried out by acting experts (**Table 1**).

6. Conclusion and recommendation

6.1 Conclusion

It can be concluded that both the refugees and SPLA soldiers had profound negative impacts on wildlife resources in the park and the local method of hunting was changed to hunting with automatic weapon, they purchased from SPLA soldiers which become one of the key management challenges for the park and also seven South Sudanese Refugee's Camps are located at three woredas of Gambella Regional State. The other management challenges for the park were previous and current resettlement programs that brought adverse impacts on vegetation and wild animal population. Some villages were wrongly planned and placed inside the National park. The government villgalization program has not been done in plan way that considered the presence of Gambella National Park from 5061 km^2 to 4,373 km^2. It can also be concluded that the institutional change result in serious turnover in staff and lack of proper care for the official documents.

6.2 Recommendation

- As there is increasing deforestation by refugees, restoration of ecology, and/or area closure should be done on surrounding areas of the camps.
- The UNHCR should give compensation in the form of funds for reforestation projects in the region in general and for the park in particular.
- As hunting and poaching are challenges for wildlife management of the park, disarmament should be done for refugees and local community by the government.
- As the resettlement program and agricultural investment are key management challenges for the park, integrating land use planning and consultation for the development program should be done in the region.

Acknowledgements

I would like to sincerely express my thanks and gratitude to Regional, Woreda key informants and field assistants. Without their information support, this research could not be very fruitful and thanks go to Gambella University for funding this research and institutions which are members of Gambella Regional State investment board for providing a relevant data for this study. I would also like to thank the UNHCR senior staff in particular Mr. Kwandwo Fempong for providing me data on Refugee population of Gambella Region. I would also like to thank experts of Gambella National Park for their assistant on the information regarding the management challenges of Gambella National Park. Last but not the least goes to Gambella National Park Scouts for their assistance during the data collection, in particular, Ato Ochan Both, Ato Ojulu Ogula Ato Mony Kong, Ato Yuoal Puoch, Ato Nyikew Obong, and Ato Top Gach for their tireless effort during the data collection. I would also like to thank individuals W/r Momina executive secretary of Research, Community Services, and industry linkages Vice President for her printing support throughout my study.

Author details

Gatluak Gatkoth Rolkier[1*] and Mulugeta Ruot Kuon[2]

1 Wildlife and Eco-tourism Department, College of Agriculture and Natural Resources, Gambella University, Ethiopia

2 Faculty of Public Administrations, Gambella University, Ethiopia

*Address all correspondencc to: grolkier2018@gmail.com

References

[1] Nelson JG, Sarafin R. Contributions of National Parks and Protected Areas to Heritage Conservation, Tourism and Sustainable Development. In: Nelson JG, Serafin R, editors. National Parks and Protected Areas: Keystones to Conservation and Sustainable Development. *Proceedings*, NATO Advanced Research Workshop on contributions of National Parks and Protected Areas to Heritage Conservation, Tourism and Sustainable Development. Krakov, Poland; 1997. pp. 1-10

[2] Searle R. Phantom Parks: The struggle to save Canada's National Parks. Canada: Key Porter Books; 2000

[3] Machlis G, Soukup M. Usable knowledge for national park and protected area management: A social science perspective. In: Nelson JG, Serafin R, editors. National Parks and Protected Areas: Keystones to Conservation and Sustainable Development. Proceedings. 1997

[4] Nathaly H. The Challenge of National Park Management A comparison of Management Plans of the Blue Mountains and Banff National Parks in accordance with Ecological Sustainable Development. Thesis for the fulfillment of the International Master's in Environmental Science Lund, Sweden National Parks of the World. Italy: White Star S.r.l; 2004. pp. 244-254

[5] Ngorongoro Conservation Area Authority (NCAA). Ecological Monitoring and Research Monthly Report. USA: The University of Arizona Press; 2002

[6] Getachew M, Weldemariam T. Review of key wildlife threats factors from literature and observation perspectives: A way forward for sustainable wildlife genetic resource conservation practices in Ethiopia. The Journal of Zoology Studies. 2016;**3**:1-12

[7] Duffy R, St John FA, Büscher B, et al. Toward a new understanding of the links between poverty and illegal wildlife hunting. Conservation Biology. 2016;**30**:14-22

[8] EWCO. Ethiopian Wildlife Conservation Organization. Compendium of Information on Individual Wildlife Conservation Areas, Volume II; 1993